NASA and the Space Industry

New Series in NASA History

ROGER D. LAUNIUS, *Series Editor*

Before Lift-off: The Making of a Space Shuttle Crew
by Henry S. F. Cooper, Jr.

The Space Station Decision:
Incremental Politics and Technological Choice
by Howard E. McCurdy

Exploring the Sun: Solar Science since Galileo
by Karl Hufbauer

Inside NASA: High Technology and Organizational Change
in the U.S. Space Program
by Howard E. McCurdy

Powering Apollo: James E. Webb of NASA
by W. Henry Lambright

NASA

AND THE

SPACE

INDUSTRY

▼

Joan Lisa Bromberg

JOHNS HOPKINS UNIVERSITY PRESS
Baltimore and London

©1999 The Johns Hopkins University Press
All rights reserved. Published 1999
Printed in the United States of America on acid-free paper

Johns Hopkins Paperbacks edition, 2000
2 4 6 8 9 7 5 3 1

The Johns Hopkins University Press
2715 North Charles Street
Baltimore, Maryland 21218-4363
www.press.jhu.edu

Library of Congress Cataloging-in-Publication Data will be found
at the end of this book.
A catalog record for this book is available from the British Library.

ISBN 0-8018-6532-8 (pbk.)

to my dance teachers, who kept me aligned

Contents

Preface

I WROTE THIS BOOK FROM AUGUST 1993 TO FEBRUARY 1998 under contracts to the National Aeronautics and Space Administration's History Office. It was my first venture into space history and I was delighted to see how many appealing questions turned up in the material. They fall under rubrics as diverse as the locus of innovation, the role of agency self-interest in the formulation of industrial policy, and ideology as a factor in determining the structure (that is, the mix of large and small firms) of industries. Many of these topics are well known to members of the space business community. My object in this book is to place them before the audience of historians of business and technology, embedding them in a narrative of NASA-industry relations in the half century since World War II.

Because my mandate was to cover fifty years, my strategy was to rely mainly on published sources, chief among them, secondary literature, trade and business magazines, and congressional hearings. I discovered, however, that for many of the episodes I treat here, the monograph literature is inadequate, and a systematic study of manuscript sources is badly needed. I hope that this book will stimulate scholars to further research, so that the next synoptic treatment of NASA's relations with industry may be more authoritative.

AT TIMES it seems extraordinary that so many people will give of their time and energy to help a fellow being write a book. I should like, first of all, to thank the members of the aerospace community who so generously granted me interviews and furnished me with material. Many of their insights will be found in the pages that follow.

I am deeply obliged to the experts who read all or parts of prior drafts and pointed out some of their shortcomings. Among them were J. Leland Atwood, William H. Becker, Roger E. Bilstein, Andrew J. Butrica, Daniel P. Byrnes, Jonathan Coopersmith, Dwayne A. Day, Henry C. Dethloff, J. D. Hunley, John W. Mauer, Diana Davids Olien, Alex Roland, David J. Whalen, and Albert D. Wheelon. I have benefitted greatly from their criticisms and suggestions.

I should like also to acknowledge the historians, records managers, and archivists who have helped me with advice or materials. They include Glenn E. Bugos, Martin J. Collins, Virginia P. Dawson, Alan Grady, Thomas R. Honan, Janet Kovacevich, Jacob Neufeld, David S.F. Portree, Lee D. Saegesser, Marion Sellers, Patricia D. Sinclair, and Michael Wright.

I owe a profound debt to the History and Philosophy of Science Program at the University of California at Davis for providing me with an intellectual home. I thank my brother, David H. Bromberg, for editorial work and for putting the manuscript on the computer. Finally, I would like to express my gratitude to Roger D. Launius, NASA's Chief Historian. He helped me in every way he could with this history, while giving me complete freedom to carry it out as I saw best.

Although the money came from NASA and substantial aid came from the aerospace and history communities, the views are mine and I alone am responsible for the errors.

NASA and the Space Industry

I

Partners in Space

N ASA, THE NATIONAL AERONAUTICS AND SPACE ADMIN-
istration, was established in 1958 as an agency of the administrative branch of the U.S. federal government. The U.S. space
industry, the other term in the title of this book, consists of a group of
large and small private firms that produce, as least part of their output,
launchers, satellites, rocket engines, and other kinds of space hardware
and services. This group began to take shape during and after World War
II as rockets and then satellites entered the military arsenal.

These two entities, one public and the other private, have been closely
associated from the start. Leaders of the aerospace industry (a larger category of firms that deal in both aircraft and spacecraft) were among the
Wise Men who in 1958 gave counsel on NASA's conception and birth.
NASA has always procured research and artifacts from the space industry. It staffs its advisory boards with space industry members and recruits
its administrators from among them. NASA and industry have joined in
petitioning Congress and successive administrations for costly space projects. They have fought with each other over patent rights and NASA contract decisions. They have even inserted themselves into each other's internal affairs, as, for example, when the industry seeks to influence NASA's
choice of programs or NASA tries to alter a company's organization or its
personnel decisions.

Of the several issues that could be raised about the NASA–space industry relation, this book discusses two. The first concerns the innovation process. In the production of novel artifacts for NASA's use by companies under contract, how did NASA and its contractors divide up activities? Who did the design? Who determined the fabrication process? What expertise did the public and private sectors bring to the table and how was it transferred between them? Who integrated the subsystems into the final articles? And, because this book is a history, how did the division of labor between the public and private sectors in this innovation process change over the decades?

The second issue addressed in this book has to do with the creation of commercial space industries, like the communications satellite industry or the launch industry. What role did NASA have in the emergence of these new sectors? Who defined their research and development (R&D) needs and who carried out the R&D? When did the private sector assume responsibility for the financing and when did it request (and secure) help from government? When were attempts made to incorporate private, profit-making elements into public, taxpayer-financed projects, or to design systems for joint public and private use? Under what circumstances did NASA abet the formation of commercial space enterprises and when did it act in the other direction, attempting to thwart them?

The book thus unites a study of NASA's relations with its contractors with a study of its relations with commercial space firms. These two issues belong together, first of all, because NASA from the start has been an agency with two missions.[1] When it was formed in 1958, it inherited the personnel and facilities of the National Advisory Committee for Aeronautics (NACA), and with them it acquired the NACA's mission of research and development in support of commercial industry—in the NACA's case, the aircraft industry. The exhilarating projects mandated by the cold war were superimposed on this inherited charge. Missions like the Mercury project or the Apollo program had little to do with commercial payoffs. They served rather to demonstrate American power, to restore the nation's self-esteem, and to raise its international standing.[2] But while these projects held the spotlight, the old NACA mission in aid of industry continued and spread from aeronautics to space.

The two issues also belong together because no rigid wall separates the contracting firms from the commercializers. On the contrary, many companies have played both roles. Some firms used NASA contracts to put themselves in a position to manufacture commercial space systems, while others mounted internally funded commercial projects primarily to make themselves more attractive candidates for NASA contracts. Recently, NASA asked industry for hardware that could serve the agency for experiments at the same time that it provided participating companies with a waystation toward commercial products. Any history of NASA-industry relations that treats NASA's contractors is thus obliged to look at the agency's interactions with commercial space and vice versa.

I TAKE a slightly different cut through these issues with the affirmation that, in NASA's case, public-sector participation in technological innovation and in the emergence or maintenance of industries are inextricably intertwined. These two topics, by and large, have been treated in disparate literatures. Technological innovation and the manner in which distinct organizations interact to produce it has been mainly the province of historians of technology and business. Their interest has been wider than government-private interactions. The cases they have examined have ranged from how manufacturers of capital goods cooperated in invention with the firms they supplied, to how companies collaborated in product design across national boundaries, to how innovations were shaped jointly by industrial firms and ordinary consumers.[3]

Innovations involving government and industry have been a special case of these studies. In particular, while economists debate whether innovation has more often been carried out by large firms or small ones, business and technology historians have examined what may be a more typical case, in which innovation is the product of networks of government and private organizations.[4] Historians of technology have also contributed the insight, to which I shall return shortly, that innovators are at one and the same time members of an organization—be it firm or public agency—and of a wider community of specialists. The General Electric physicist-inventor is part of GE but also of the physics community; the Pratt & Whitney design engineer also belongs to a national community of engine practitioners.[5]

In contrast, the role of the government in the creation or sustenance of industrial sectors has mainly been studied by economists and political scientists. They have looked at the effects of government-mandated regulations on industry. They have weighed how much government bureaucrats have been ruled by conceptions of the public weal and how much they have sought to maximize the budgets, power, or reputation of their department.[6] They have shown how government procurement of R&D functions as a form of "industrial policy," defined as policy that fosters particular industrial sectors for the benefit of the economy as a whole.[7]

They have pointed out that there are national styles in government intervention in the private sector, as well as different styles within a single nation for dealing with various sectors of industry.[8]

The NASA case is relevant to both these literatures and it throws light
on each of them. We shall see that it provides a consummate example of
innovation carried out by a network of private and public organizations.
At the same time, it is a case study of the means by which a public agency
has influenced the formation and sustenance of new industry.

But the NASA case has an additional advantage. Whenever two distinct circles of discourse are brought together, we may hope that new insights will be generated and new questions raised. The history of NASA-
industry relations fulfills such an expectation. The bureaucrats at NASA
have done more than seek to maximize the public benefits of space or their
own power or budgets. They are engineers, scientists, or science managers
as well as bureaucrats, and so seek to maximize quantities like the amount
of interesting research under their purview or the amount of technical
expertise within their departments. As technical professionals, they have
been immersed in the same wider community of practice that has bathed
their counterparts in industry. Certainly, the interactions between NASA
and industry have been shaped by the cultures and interests of the agency
on the one hand and the several firms on the other. But they have also,
often, been mediated by the loyalty of both NASA and industry personnel to aerospace engineering.

The NASA case also links the two literatures by showing that public
agencies affect the private sector through more than regulations, procurements, or R&D funding. NASA helped to determine the very shape
of the innovations that lay at the basis of some industries and the trajectory the technology would take.

It may be objected that NASA is a special case, that space technology
and the space industry have characteristics that have made a large measure of government participation inevitable. Space has been intimately
connected with U.S. foreign policy. In the 1960s it was enlisted in the service of the cold war; in the 1990s the space station became woven into
U.S.-Russian relations. The space industry also has some special claims
on government financial help because it has had to insure itself against the
devastation that a fallen spaceship could inflict on a foreign nation. In addition, space technology has a military usefulness that has never abated.
For the first thirteen years after World War II, space in the United States
mainly meant missiles. Space science piggybacked on military projects.[9]
Even after NASA was created as a civilian space agency, missiles retained

their importance, while military reconnaissance, communications, navigation, and weather satellites became steadily more essential.

Furthermore, the space industry overlaps the defense industry, and the point has often been made that the defense industry cannot truly be considered a "private" sector.[10] It serves a single consumer. It often receives "cost plus" contracts, which pay a fee plus all incurred expenses, and thus free companies from the normal market discipline of keeping costs low.

These caveats, however, overlook far-reaching analogies between the space industry and other sectors. Solid-state electronics, masers and lasers, and nuclear technology all started under the wing of the military.[11] Commercial nuclear power was an instrument of U.S. foreign policy in its time, and there are interesting parallels between the way in which the U.S. Atomic Energy Commission tried to foster commercial nuclear energy and NASA's promotion of space industrialization.[12] Finally, regardless of the special circumstances that clothe sales to the government, deep differences remain between private companies and public agencies, if only because private companies direct so much of their actions toward maximizing profits. In short, space is not so peculiar that it cannot teach us something about both technology-creation and industry-creation.

THIS BOOK starts with the formation of the U.S. space industry in the 1940s and ends in the early 1990s. The story it tells, stretching over half a century, played out against a changing economic and political background, as well as a changing appreciation of the role of the public vis-à-vis the private sector. The United States emerged from the Second World War with its productive capacity enlarged, whereas the other major participants' economies were in chaos. Throughout the 1950s and the 1960s, the United States remained the hegemonic economic power. U.S. firms dominated the markets of the capitalist world and led the world in technology. Economic prosperity was one of the factors that enabled the United States to sustain a vigorous cold war against the Soviet Union.

But a period of economic malaise set in in the late 1960s and early 1970s. By 1970, Europe and Japan had recovered from the devastation of the world war, and their industries began to pose a threat to U.S. firms in both domestic and overseas markets. Inflation grew and the dollar weakened. The federal budget was strained by the Vietnam war and the social programs of the 1960s. The U.S. public shifted its interest from the cold war to the economy, the environment, and the plight of the cities. Economic conditions worsened through the 1970s. The United States and the industrialized world sank into a condition of low growth, high unemployment, and high inflation that lasted through the early 1980s.[13]

The changing economic conditions were reflected in changing opinions on the proper roles of government and industry. During the prosperous 1950s and 1960s there was widespread agreement that the government should take measures to keep the economy robust and unemployment low. But two contrasting positions emerged as the United States entered the troubled 1970s. One position championed industrial policy, that is, government intervention to support those industries liable to provide overall economic benefits. Opposing it was the position that government should intervene in the economy as little as possible. Adherents advocated deregulation of industry, a cutback in the amount of government R&D on behalf of the private sector, and privatization of state and federal government functions. This point of view became administration policy in the 1980s, during the presidencies of Ronald Reagan and George Bush.[14] It continued to be a dominant strain in the first part of the 1990s even though the U.S. economy improved. The ideology of shifting government functions to the private sector was one of the factors that drove a veritable frenzy of federal budget cutting in the early 1990s.

NASA itself changed over the years, partially in response to these broader transformations. To begin with, NASA, as a federal agency, was quite rightly viewed by successive Congresses and administrations as an instrument of national policy. New federal policies and priorities were therefore bound to affect the agency. In the 1960s, NASA served as a means for waging the cold war. Once Richard M. Nixon took over the presidency in 1969, however, this role became less necessary. Nixon moved toward détente with the Communist world, so that space as an arena for competing with the Soviets or for garnering international prestige lost importance. It was the state of the economy that was now the preoccupation. At the same time, space commerce was becoming a reality as after years of research and planning, communications satellites began to enter the market. In this climate, administrations and Congresses more and more harked back to the old National Advisory Committee for Aeronautics and its service to the aircraft industry, or to NASA's own research on communications satellites. They called on NASA to take on the role of auxiliary to commercial space. NASA responded by increasingly justifying its core programs as contributions to the commercialization of space.

Economic factors also came into play. Programs like Apollo had been predicated on a vision of the United States as a nation with unlimited resources, one that could do anything it set its mind to.[15] In contrast, during the 1970s and after fiscal restraint was in order. There were no more funds for space spectaculars like the moon landings. NASA was constrained to show that its programs had an economic payoff.

A vital entry in any catalog of changes must be NASA's shifting relations with the military services.[16] The services, after all, preceded NASA as the federal agency that bought space R&D and hardware. By the time NASA was created in 1958, the Department of Defense (DoD) and its three arms had had nearly two decades of experience with a space industry that, for its part, had developed in tandem with the military market. After 1958, NASA and the Defense Department became the two major government bureaus purchasing space goods. At times, as in the 1960s, NASA was the greater spender; at other times, as in the 1980s, the military pulled ahead. For industry, the federal government was a bipartite market in which the relative importance of the two parts kept changing.

Equally significant was the changing partition of missions between the civilian and military sides of government. A tug-of-war over who would have what role went on for more than a decade. At the start of the 1960s, NASA won from the military the right to expand into research on geosynchronous communications satellites. At the end of the 1960s, the Air Force lost jurisdiction over systems flying astronauts to NASA. Conversely, during the mid-1980s while NASA struggled with its shuttle launch vehicle, the Department of Defense became the main government procurer of expendable boosters. Thus it was not only relative budgets that changed, but the areas in which the two agencies could offer contracts.

Along with such changes went technological advances. Communications satellites were developed that could link nations over the Atlantic and Pacific oceans. Further improvements permitted communication over smaller areas, like regions or single nations. Surveillance satellites achieved finer and finer resolution. This was useful for their initial purposes—military and intelligence tasks—but it also offered commercial possibilities. The same was true for navigation satellites; they began as security systems but had potential for civilian navigation. A technology that could sustain a nongovernment market emerged.

I have introduced these currents—political, economic, ideological, institutional, technological—as "background," but in fact they were the forces that drove NASA's relations with industry. It is true that there were also what might be called "internal" factors. Over the decades, NASA and its contractors learned to work together ever more smoothly. They framed better procedures for making decisions and for documenting the configuration and the fabrication processes for the artifacts they were building together. NASA honed its contracting methods to get better performance from industry and retain more control. By and large, however, the dynamics that governed how NASA and industry shared the innovation process or how NASA formulated its policies toward commercial space did not come from such changes, but from the outside. In the narrative that follows, I pay some attention to internal developments, but chiefly explore the links between larger political, economic, and ideological events and the NASA-industry nexus.

NASA is not a monolith. It is composed of a Washington, D.C., headquarters, which sets policy and handles exterior relations, and field centers scattered from Florida to California. Chart 1 shows the locations and provenance of major NASA centers. Centers and headquarters all work together, but they can also be at loggerheads. Centers compete among themselves for allocation of resources or scrap over which of them will secure a given project. Centers and headquarters often fight over which should have decision-making power in a given project.

The "space industry," of course, must also be disaggregated. It is composed of individual firms that are perpetually entering the space sector, merging with other firms, leaving the field, and so on. Chart 2 summarizes the histories of some of the companies discussed in this book. These firms sometimes cooperate on a project, as one subcontractor to another or as associate contractors. Often, they compete viciously for government contracts or shares of commercial markets. Or they try to kill government programs that will give a competitor an advantage and advance others that will help themselves.

Struggles occur, not only within NASA or the space industry but between NASA units and private firms. Centers sometimes try to retain in-house programs that firms would like to see contracted out. Or, in a particularly relevant example for our purposes, private companies and NASA units compete for tasks that will increase their expertise. A firm that can keep design, systems engineering, and systems integration in-house maintains the highly skilled engineers it needs to win new contracts. It garners prestige for being innovative and capable of difficult engineering, which again translates into better chances for contracts. It gains the opportunity to structure the hardware it is building to make it more broadly marketable, as when a company building a plane for the Navy uses its control of design to fashion the plane so that it can later be adapted for sale to the Air Force.[17]

For its part, government also has reasons for keeping design and systems engineering in its province. These are some of the more interesting jobs in engineering, and retaining them in-house helps government agencies attract good people. Technically competent personnel, in turn, allow the government to oversee its industry contracts properly. Prestige is as valuable a commodity in government agencies as it is in private companies. The more prestige accrues to it, the more an agency is able to pry funding and new projects out of Congress and the administration. Government, furthermore, is well aware that industry designs can embody company self-interest, and may feel the need to forestall such designs. When NASA and industry fight over who performs the tasks that confer expertise, they are, of course, fighting over how to divide innovation activities. This is one of the ways in which organizational self-interest exhibits itself within the matter of the public-private division of labor.

Of course, there is cooperation as well as conflict in the NASA-industry relation. We have mentioned that NASA and industry engineers are united by their common participation in and loyalty to the practice of space engineering. The two groups often shared a common version of patriotism—a commitment to winning the space race in the name of the United States and the capitalist system in the 1960s, and in the 1970s and after a commitment to building the competitiveness of U.S. industry. Up and down the line, NASA personnel and the personnel of its contractors shared a dedication to the success of projects like Apollo, the shuttle, and the space station. In general, they strove to design and fabricate the best possible spaceware. And when the work was done, they shared their pride for what have often been substantial accomplishments.

NASA, industry leaders, and congressional members who supervise the space program like to talk about a public-private partnership. The 1969 moon landing was the result of "a highly productive partnership . . . among government, industry, and the university community."[18] "What we hope to do . . . is . . . to encourage private industry to form the same kind of partnership that NASA has had with the aeronautics industry."[19] "What is at stake is whether the Government will enter into a fair risk-sharing arrangement with the private sector . . . to enable the private sector to work as a partner with the Government in the continued development of space."[20] "The goal . . . is to establish partnerships among the government, the private sector, and academic institutions to facilitate . . . the commercialization of space."[21] Solving the problem of cheap and effective transportation into orbit "is a public-private partnership responsibility" and the NASA program to develop such access is "a government-industry partnership."[22]

To be sure, most such pronouncements have not been just expressions of warmth and coziness. Rather, they have been rhetoric in service of attempts to gain some specific result, from the performance of research and development on behalf of commercial industry to financial arrangements

MARSHALL SPACE FLIGHT CNTR
Huntsville, Alabama

| Army Ordnance Guided Missile Center (Est. 1950) | Becomes part of the new Army Ballistic Missile Agency, 1955 |

LANGLEY RESEARCH CENTER
Hampton, Virginia

| Est. 1917 as Langley Memorial Aeronautical Laboratory of NACA | Becomes NASA Langley Research Center, 1958 | 1958 Space Task Group formed at Langley |

JOHNSON SPACE CENTER
Houston, Texas

GODDARD SPACE FLIGHT CNTR
Greenbelt, Maryland

| Est. as a NASA Center in 1959 |

KENNEDY SPACE CENTER
Brevard County, Florida

| Est. 1949–1950 as an Air Force Missile Launch Range | ABMA establishes its Missile Firing Laboratory There |

LEWIS RESEARCH CENTER
Cleveland, Ohio

| Est. 1941 as NACA Aircraft Engine Research Laboratory. Renamed Lewis Flight Propulsion Laboratory, 1948 | Becomes NASA Lewis Research Center, 1958 |

AMES RESEARCH CENTER
Moffett Field, California

| Est. 1940 as NACA Ames Aeronautical Laboratory | Becomes NASA Ames Research Center, 1958 |

JET PROPULSION LABORATORY
Pasadena, California

| Jet Propulsion Laboratory est. 1944. Run by California Institute of Technology under contract to U.S. Army | JPL Facilities and contract transferred to NASA, 1959 |

1950   1960

Some Major NASA Installations. The centers of NASA were made up of NACA installations, laboratories ceded by the Armed Services, and a facility, the Goddard Space Flight Center, built specifically for the new agency.

Development Operations Division led by von Braun shifted from ABMA and becomes NASA's Marshall Space Flight Center, 1960

Space Task Group becomes Manned Spacecraft Center, relocated to Houston, 1961–1964

Renamed Johnson Space Center, 1973

Made Independent of MSFC as Launch Operations Center, 1962

Renamed Launch Operations Directorate of NASA's Marshall Space Flight Center, 1960

Renamed Kennedy Space Center, 1963

1960   1970   1980

Sources: Jane Van Nimmen and Leonard C. Bruno with Robert L. Rosholt, *NASA Historical Data Book,* Vol. 1, *NASA Resources, 1958–1968.* Washington D.C.: NASA, 1988; Ihor Gawdiak with Helen Fedor, *NASA Historical Data Book,* Vol. 4, *NASA Resources 1968–1978,* Washington D.C.: NASA, 1994; and Roger E. Bilstein, *Orders of Magnitude: A History of the NACA and NASA, 1915–1990,* Washington D.C.: NASA, 1989.

GENERAL ELECTRIC

Founded 1892

THIOKOL CHEMICAL Co.

Founded 1928

REACTION MOTORS

Founded 1941

AEROJET ENGINEERING Co.

Founded 1942
Becomes part of General Rubber
and Tire in 1944

NORTH AMERICAN AVIATION

Founded as Holding Company   Becomes aircraft manufacturer
1928   1935

BOEING AIRCRAFT Co.

Founded 1916

DOUGLAS AIRCRAFT Co.

Founded 1920–1926

McDONNELL AIRCRAFT Corp.

Founded 1939

LOCKHEED BROTHERS   **LOCKHEED AIRCRAFT Co.**

Aircraft Builders   Founded 1926

GLENN L. MARTIN Co.

Founded 1912–1917

HUGHES AIRCRAFT COMPANY

Founded as division of Hughes Tool, ca 1932

| 1900 | 1910 | 1920 | 1930 | 1940 | 1950 |

Some U.S. Companies in the Space Business. Aircraft and electric companies formed before World War II diversified into space hardware after it. In the 1950s and 1960s, some shared in the widespread strategy of conglomeration with unrelated businesses; in the 1990s, mergers and acquisitions sharply reduced the number of firms.

GE buys RCA Corp., 1986

GE sells its Aerospace
Division to Martin Marietta,
1993

ORBITAL SCIENCES

Founded 1982

1958
Bought by
Thiokol

Merges with Rockwell Standard   Boeing buys

to form NORTH AMERICAN ROCKWELL   Rockwell Aerospace
1967   and Defense, 1996

McDONNELL DOUGLAS Corp.   McDONNELL DOUGLAS

1967 Merger   merges into BOEING
1996–1997

LOCKHEED–MARTIN

Merger, 1994

Merges with American Marietta   Martin Marietta acquires

to form **MARTIN MARIETTA,**   General Dynamics' Launch
1961   Vehicle line, 1993

HUGHES AIRCRAFT COMPANY

Established as independent company, 1961   HAC becomes unit of General Motors Corp., 1985

RAMO WOOLDRIDGE

(Spinoff, 1953)   Merges with Thompson to form TRW, 1958

1950	1960	1970	1980	1990	2000

Sources: John B. Rae, *Climb to Greatness: The American Aircraft Industry,
1920–1960,* Cambridge Mass.: MIT Press, 1968; Donald L. Barlett and James B.
Steele, *Empire: The Life, Legend, and Madness of Howard Hughes,* New York:
W. W. Norton, 1979; and *The Wall Street Journal Index;* various years.

in aid of particular firms, to the reverse, that is securing industry funding for NASA programs. Still, if we accord full recognition to the fact that partnerships, be they domestic, business, or government-private, involve contention as well as concord, and that a phrase can be ironic and descriptive at the same time, then it is not unreasonable to call NASA and industry, as this chapter does, "Partners in Space."

I HAVE divided the narrative into six periods and devoted a chapter to each. The first, from 1940 to 1960, is the subject of chapter 2. It saw the formation of the U.S. space industry, and, in 1958, of NASA as well. NASA was born into the cross-fire of fights over public-private roles in space. The Air Force, which gave its prime contractors technical direction over the weapons systems they were building, had deviated from that practice in 1954 in order to build intercontinental ballistic missiles. It had inserted a newly organized firm, Ramo-Wooldridge, as intermediary between itself and its other contractors and charged the firm with technical direction and systems engineering. The aircraft companies that were the Air Force's traditional contractors vigorously disputed the change. They also attacked the Army's arsenal system, in which Army laboratories designed missiles in-house and manufactured the first units, before farming out the production models to industry.

These controversies were the more acrimonious in that the space industry was just taking shape, with aircraft companies, electronics firms, and chemical companies all struggling for hegemony over the new field. NASA itself was put together from a number of separate organizations and research teams, each with its own tradition of dealing with private companies. The legacies of arguments over Air Force and Army practices, and the habits of NASA's predecessor organizations were part of the background as NASA tried to define a suitable relation to industry in its first years.

The second period, treated in chapter 3, is the eight years that began with John F. Kennedy's decisions to promote Apollo into a crash program and accelerate the development of communication satellites. I lay out historian David J. Whalen's argument that NASA's participation in communications satellite research helped determine the trajectory of the satellite industry. I then proceed to the story of how NASA and its largest Apollo contractor, North American Rockwell, learned to work together.

The 1970s are a new period for by 1969, NASA lost the status to which Kennedy's decisions had elevated it. In chapter 4 I contend that this loss of political clout helped transform the design of spacecraft. The Mercury capsule and the Apollo stack had been designed in-house by NASA engineers. With the 1970s space shuttle, design became encumbered by NASA's need to fend off opponents and rally supporters. The result was a process much more permeable to industry ideas.

The political climate of the 1970s made it expedient for NASA to reconceptualize the shuttle: from part of the grand adventure of human mastery of space, it became a cost-effective transportation system. In the rest of chapter 4, I examine the impact that had on space commerce. Specifically, I look at the effects on the expendable launch industry, the communications satellite industry, and that amorphous group of visions, projects, and companies known as "space industrialization."

President Ronald Reagan's first term deserves its own period for its celebration of private enterprise. In chapter 5 I show how a flock of space companies were spawned by this free-market ideology, the success of the shuttle, and the very visible boom enjoyed by the commercial communications satellite business in the 1970s. To meet the needs of those businesses and to address the prodding of Congress and the administration, new offices were set up in NASA (the Office of Commercial Programs) and other agencies. But while the Office of Commercial Programs labored to facilitate the entrance into space of commercial companies, the NASA hierarchy worked to prevent their entrance into one particular field—the provision of launch services. The point is that NASA set prices to protect the shuttle program, that is, to retain as business for the shuttle the launch of private sector and foreign government satellites. I also look at NASA's return to research on communications satellites, which in contrast to the 1960s took place in the context of an established communications satellite sector.

The explosion of *Challenger* removed the space shuttle as a competitor for commercial cargoes and gave a foothold to the U.S. launch industry. This marked a new period in the history of NASA-industry relations, one treated in chapter 6. I also consider here the space station, NASA's new core program and an attempt to fashion from the beginning a system that could simultaneously meet the needs of the commercial space industry and the government.

The final period, starting about 1991, saw the cold war over, the industry contracting, and NASA policies changing under a vigorous new administrator. Because it is a period we are still in, I cast it as an epilogue, chapter 7, and use it to throw the earlier stages into relief.

Legacies

THE U.S. SPACE INDUSTRY FORMED AS A CONSEQUENCE of the rise of a market for military missiles and, toward the mid-1950s, satellites. The defense industry is often called a *monopsonic,* or single buyer, market. In fact, however, there was more than one buyer in this market in the 1940s and 1950s, because the different missile agencies in the three services—Army, Navy, and, after 1947, Air Force—adopted different modes of dealing with their suppliers. Agencies in the Army adopted a modified "arsenal" practice, so-called after the federally owned and operated arsenals of the nineteenth century, where rifles and other arms were produced. The Army missile agencies similarly did much of their design and development in-house, and sometimes even manufactured and tested the first prototypes before they let a contract to an industrial company for the large number of vehicles that would be needed for further tests and operations.

The Air Force, in contrast, developed the "weapons system" method in which a private, prime contractor accomplished design, development, and prototype construction and built the production models. A variant of this system gave design and systems engineering to a separate private firm, interposed between the Air Force and the manufacturers.

In addition to forming itself around a new market, the space industry developed around a new technology that was characterized, above all, by its multidisciplinary character. The creation of spaceware transcended mere aeronautical, chemical, or electronics engineering in that it required all these specialties and more. The corollary is that the space industry transcended existing industrial sectors. Established firms that wished to move into this new field had, somehow, to acquire additional skills.[1] And firms from different sectors did move in, colliding with each other as they did so and occasioning open jurisdictional disputes.

The manner in which the military agencies structured their buying had an impact on the interfirm and intersectorial rivalries. The Air Force system could advantage some firms and sectors and disadvantage others. So could the "arsenal" system. The pattern of contractual relations that a service followed was therefore of vital interest to companies, and disputes over these patterns were the more intense because of the jousting for control of the new space arena.

NASA was empowered to carry out the design, development, and testing of aeronautical and space vehicles.[2] It was expected to spend the bulk of its money on contracts with industry. As NASA moved into action, it was clear that its contractual relation with industry was one of the policies it would have to hammer out. Contemporaries saw NASA's early policies toward industry within the context of the policies established (and the controversies engendered) by the armed services. We need to keep this context before us also.

Another part of NASA's inheritance was the traditions of the organizations that made it up. The nucleus of NASA was the NACA, an organization that had seldom designed specific vehicles but instead had excelled in generic research, applicable to broad classes of aircraft. The NACA spent most of its money in-house and had little experience with industrial contractors.

From 1958 to 1960, NASA absorbed the Jet Propulsion Laboratory (run for the Army through a contract with the California Institute of Technology) and a part of the Army Ballistic Missile Agency. Unlike the NACA, these organizations had done extensive vehicle design and had contracted with industry for large and complex projects, although within the Army arsenal tradition. NASA also took in the group of scientists at the Naval Research Laboratory (NRL) who were designing Vanguard, a satellite vehicle.

The final ingredient in NASA's inheritance was the set of beliefs about industry-government relations held by President Dwight D. Eisenhower and the man Eisenhower appointed as first head of the agency, T. Keith Glennan. Both men espoused a philosophy of small government and of giving the maximum possible scope to the private sector. The legacy NASA was handed at birth, therefore, was one of controversy and conflicting views on the public-private partnership.

Formation of a Space Industry

During World War II rockets became a significant part of U.S. weaponry. Rocket-powered guns were built for tanks, trucks, and ships. The shoulder-mounted rocket called the bazooka was invented for infantrymen. Rocket missiles were developed for use by airplanes against ground and air targets, and, conversely, rocket-powered antiaircraft guns were developed to shoot down planes. At the Guggenheim Aeronautical Laboratory of the California Institute of Technology in Pasadena, California, and at the Navy's Allegany Ballistics Laboratory in West Virginia, engineering work was carried out on jet-assisted take-off (JATO) rockets that could provide extra force for the take-off of heavily laden military planes.

To manufacture these weapons, the federal government gave contracts to a variety of firms. Some were established companies. The Hercules Powder Company, a chemical firm dating to before World War I, became the leading wartime producer of solid rocket propellants. General Electric was commissioned to develop rocket fuses. Bell Telephone Laboratories, working through Western Electric Company, helped with engineering.[3] Other companies were created for the work. Four members of the American Rocket Society formed Reaction Motors Incorporated in Pompton Plains, New Jersey, at the end of 1941, after they secured a Navy contract to work on a rocket engine of their own design. Some California Institute of Technology scientists, together with their attorney, formed Aerojet Engineering Corporation in early 1942, to manufacture the JATO engines the scientists were designing.[4]

"Rocket" is not necessarily the same as "space system." Rockets are projectiles that house within themselves both fuel and the oxidizer needed to burn the fuel. They derive their propulsive force from the ejection of the products of the combustion. (The laws of mechanics tell us that the rocket receives a forward momentum of the same magnitude as the backward momentum of the combustion gases.) The rockets that the United States used in World War II never left the atmosphere. But rocket technology is fundamental to the launchers that do reach space or send craft into it, and companies that designed or manufactured the wartime rockets were gaining an insight into space technology.

In mid-1943, word reached the United States that the Germans appeared to be developing rockets (the infamous V-2s) of longer range and heavier payload than anything the Allies had. In response, the U.S. War

Department expanded its rocket R&D. Among the new projects undertaken, Army Ordnance contracted with General Electric for wide-ranging rocket research under Project Hermes. AT&T received a contract for Nike, an antiaircraft rocket. Army Ordnance also signed a contract with the California Institute of Technology for a new laboratory, the Jet Propulsion Laboratory, to be devoted to all aspects of guided missile research and development. It would be funded by the Army and run by the institute.[5]

When the war in Europe ended, the U.S. Army seized and brought to the White Sands Proving Grounds in New Mexico hundreds of parts of V-2 rockets. It expanded the Hermes contract to allow General Electric to supervise the reassembly and launch of the Nazi rockets. Army Ordnance also brought back another kind of booty, 127 German V-2 scientists, to work on the rockets out of a base close to White Sands—Fort Bliss in Texas.[6]

Aircraft companies, in the main, did not participate in rocket production during the war. Their plates were full as they struggled to increase production from the fewer than 6,000 planes they had turned out in 1939 to the more than 96,000 they produced in 1944. At the end of the war, however, the military canceled its orders for war planes and industry revenues plummeted. Sales of U.S. aircraft manufacturers went from 16 billion dollars in 1944, the last full year of war, to 1.2 billion dollars in 1947, a more than tenfold decrease.[7]

Companies met the postwar defense slump with different strategies. Some placed their bets on a pent-up demand for small private planes. Others eyed the commercial airliner market.[8] It was clear, however, that the war had brought to the point of practical application a number of spectacular new technologies that seemed a natural extension of the aircraft companies' core business—atomic energy, which was then being considered for both aircraft and rocket propulsion, radar, jet propulsion, and rocketry.[9]

One company that moved aggressively in using the assets accumulated in wartime to diversify into these new fields was North American Aviation. North American was organized as a holding company for firms related to aviation in 1928, but had been reorganized into an aircraft manufacturer in 1935. It emerged as one of the leading manufacturers of military planes during the war, with a peak work force of nearly 90,000. Within months after the end of the conflict, its work force had dropped to 5,000.

James H. Kindelberger, a World War I pilot, and John Leland Atwood, a civil engineer turned aeronautical engineer, had led the company since its reconstitution in 1935. In framing a postwar strategy, Kindelberger and

Atwood saw no wisdom in moving into commercial aircraft as Martin and Consolidated Vultee (Convair) were attempting to do; there was too much competition from prewar leaders Douglas, Boeing, and Lockheed. They decided to stay with the military market. North American had contracts for some early jet planes from the services. Kindelberger and Atwood added to this a capability in weaponry based on the new technologies. They hired a cadre of scientists knowledgeable in electronics and rocketry, many of whom had military experience, and set up a new organization, the Aerophysics Laboratory, where the scientists might work on rocket engines, inertial guidance, atomic energy, and missiles. At the head of the group they placed Willam Bollay, a scientist who had trained in rocket research at the California Institute of Technology. In making these moves, Kindelberger and Atwood were inspired, in part, by the V-2 rockets. These weapons were significantly larger and faster than any rocket the United States had produced, and they presaged a new age in missilery.[10]

While North American was taking these steps, the Army Air Force was sponsoring nearly 30 guided missile contracts, most with aircraft firms. North American's Aerophysics Laboratory received the important Navaho cruise missile, which used both rockets and air-breathing engines.[11] The Navy also contracted for space-related projects. The Naval Research Laboratory commissioned a sounding rocket, the Viking, from the Glenn L. Martin Company, with the motor to be built by Reaction Motors. The Navy Bureau of Aeronautics sponsored a preliminary structural design of an earth-orbiting satellite, letting contracts to Martin and North American.[12] All this contributed to bringing aircraft firms into the space business.

Chemical companies were also tapped. The Hercules Powder Company was hired to take over management of the Navy's Allegany Ballistics Laboratory, a facility devoted to R&D on solid rocket motors. The Thiokol Corporation in Trenton, New Jersey, discovered that a polysulfide polymer it had invented was being systematically purchased by the Jet Propulsion Laboratory for use as an experimental solid rocket fuel. Thiokol was formed in 1928 to produce the substance "thiokol," which its inventors hoped would be a popular synthetic rubber. The rubber sales were limited, however, and the company had not prospered. Now Thiokol president James W. Crosby decided to move his small company into the solid rocket motor business. Thiokol worked with Jet Propulsion Laboratory scientists to improve its polymer and in 1947 Crosby got a contract from Army Ordnance for a pilot plant to manufacture it.[13]

Thus the "space industry" was begun. The result of wartime and postwar military procurements, the industry was made up of new entrepreneurial firms and dedicated divisions and groups within aircraft, chemical, and electronics companies.[14]

MILITARY INTEREST flagged in 1947 and 1948, as the services turned their attention to the massive reorganization that created the Department of Defense (in 1947) and separated out the Air Force as a unit coequal with the Army and Navy. Defense budgets also shrank in those years. But in 1949 and 1950, the situation changed. First, the Soviets exploded an atomic bomb in August 1949, demonstrating that the United States was no longer the sole nuclear power. Then the Truman administration embarked on a crash program to develop a hydrogen bomb. Meanwhile, Atomic Energy Commission research began to show that atomic weapons could be made small enough to be delivered by missiles, and intelligence sources reported that the Soviets had an extensive missile program underway. Finally, in June 1950 North Korea invaded the South, and the U.S. became embroiled in the Korean War. As missiles began to take on heightened importance, a furious struggle broke out among the Air Force, the Army, and the Navy over which service would win the right to field them.[15] The wrangling pushed each service into sponsoring more rocket and space projects than it otherwise would have found necessary.

The Army decided to convert the Redstone and Huntsville Arsenals in Huntsville, Alabama, which had been built in 1941 as munitions factories but idled after the war, into a missile development laboratory. In 1950, it moved the Guided Missile Development Division that had grown up around the German rocketeers from Fort Bliss to Huntsville to form the core of a new Army Ordnance Missile Center. The Center's first job was to create the Redstone, a rocket-powered ballistic missile with a range of a few hundred miles.[16]

The Air Force had emphasized cruise missiles in the lean years of the late 1940s because they could be developed more quickly than ballistic missiles. The new circumstances at the turn of the decade initiated a protracted battle among Air Force bureaucracies that resulted in a crash program to build ballistic missiles. In 1954, it approved development of the Atlas intercontinental ballistic missile (ICBM).[17] In 1955, it added a second ICBM, the Titan. By this time, the Army and Navy had begun collaboration on a liquid-propellant intermediate range ballistic missile (IRBM), the Jupiter. The Air Force hastened to initiate work on a rival IRBM, the Thor. At the same time, the Air Force started a program of reconnaissance satellites, under the designation WS (weapons system) 117L.[18] Shortly after that, in 1956, the Navy decided that liquid-propellant rockets were too dangerous to have onboard ships and submarines. It withdrew from the Jupiter project, leaving the Army Huntsville Center—by then reorganized as the Army Ballistic Missile Agency—to pursue the Jupiter by itself, and started to build the solid-fueled Polaris missile for its submarines.

Meanwhile, the world scientific community scheduled an International Geophysical Year (IGY) for July 1957 to December 1958. It called for nations to put up artificial satellites to carry out observations of natural phenomena that could not be registered on ground-based instruments because of the Earth's atmosphere. In July 1955, President Eisenhower announced that the United States would undertake to orbit such a satellite. The Army's Huntsville missile laboratory, in collaboration with the Jet Propulsion Laboratory, the Naval Research Laboratory (NRL) and, with less enthusiasm, the Air Force, all submitted proposals for this satellite and its launcher. The Army's proposal was based on the Redstone missile, the Air Force's on the Atlas, and the NRL proposed to build a new launcher, derived from its Viking rocket. The Eisenhower administration, which wished to avoid diverting the Air Force and the Army from their military projects, chose the Naval Research Laboratory's proposal, Project Vanguard.[19] Thus, as the 1950s proceeded, the U.S. armed services embarked on an ever-increasing number of missile, rocket, and satellite projects.

NOT SURPRISINGLY, this proliferation of programs stimulated a response from industry. That response, however, depended crucially on the fact that the design and manufacture of these new craft required expertise in airframe structures, chemicals, electronics, and other areas, expertise that had been housed in separate industries and now had to be brought together.

In this situation, companies adopted a variety of strategies. Some aggressively attempted to develop in-house the technologies they needed; others sought to acquire companies that already had those technologies. Some formed alliances with firms in the other industry sectors. Others bid for contracts that promised little or no profits, but would school their engineers in new skills. Many elected a combination of these tactics and almost all expanded their facilities to accommodate additional business.

As we have seen, North American Aviation's leaders, James H. Kindelberger and J. Leland Atwood, were determined to develop in-house technical and managerial competence in the new technologies. Within its Aerophysics Laboratory, established in 1946, North American Aviation had worked on the Navaho cruise missile, one of the most important of the missile projects the Army Air Forces sponsored in the immediate postwar years. North American Aviation developed its own rocket engine for the Navaho, and out of the Navaho engine it derived an engine for the Army Ballistic Missile Agency's Redstone, and so began to produce rocket engines. The electronics group it set up to engineer the Navaho's guidance systems went on to design guidance and radar for other weapons and warplanes. As these activities grew, the Aerophysics Laboratory metamorphosed into a Missiles and Control Equipment unit, and then, in 1955, into four new divisions: Rocketdyne, for rocket engines; the Missile Development Division; Atomics International, for atomic reactors, including nuclear power for missiles and satellites; and Autonetics, for electronics. At the same time, the company took pains to retain strong aircraft divisions. North American was metamorphosing from aircraft company to aerospace firm.[20]

While North American Aviation was remaking itself into an all-around aerospace company, the Glenn L. Martin Company was on a different path. This venerable aircraft producer, founded in the first decade of the century, took a trajectory toward a pure space concern. When military orders were canceled in 1945, the company's founder and president, Glenn L. Martin, gambled on entering the commercial market with two medium-range airplanes of new design, the 2-0-2 and the 4-0-4. The planes failed to garner sufficient orders from the airlines. In 1949, on the verge of bankruptcy, Glenn Martin succeeded in getting a government loan on condition he retire from the presidency to the position of chairman of the board. In 1952, with the company still on the ropes, it was the Navy that bailed it out. This time, Glenn Martin was only granted an honorary position. George Maverick Bunker, an able manager from outside the aircraft industry, was brought in as president and chief executive officer.

Meanwhile, the Martin Company got a series of contracts for both operational and experimental missiles: the Air Force Matador, the Army truck-mounted Lacrosse, the Navy Gorgon IV and Oriole test missiles. Martin had made the Viking sounding rocket for the NRL, and when that group was chosen to build the U.S. Vanguard satellite for the International Geophysical Year, Martin became the contractor because the NRL wanted the Vanguard rocket to be a derivative of the Viking.[21] In December 1955 Martin also won the contract to build the Air Force's Titan ICBM. The company now saw itself as having graduated into a "premier producer of missiles and rockets." It created a new division to handle the Titan and built a new facility in Denver, Colorado. Bunker, meanwhile, laid plans to take the company out of the airplane business altogether and have it concentrate on missiles, rockets, spacecraft, and electronics.[22]

Lockheed Aircraft Corporation was not run, in the 1950s, by Allan

Lockheed, that one of the pioneering Scottish-American Lockheed (formerly Loughead) brothers who had given the company its name. He had
withdrawn from the company before the 1929 stock market crash. In
1932, the company was taken over by a group of men who each had a different connection to the aircraft industry and Robert E. Gross, one of the
group and a former Boston investment banker, became the president.[23]

During the Korean War, Gross decided to expand the company into
missile work. Its first project was the MX-883 missile, which was not powered by a rocket but by an air breathing engine. In January 1954 Lockheed made this program the nucleus for a Missiles Systems Division located at Burbank in the Los Angeles area. It built a research laboratory
for the division and set about hiring scientists and engineers. The Lockheed Missiles Systems Division had 65 engineers at its founding; by mid-
1956, it had nearly 750. Division sales reached $7.2 million in 1954. In
1955, a year in which the corporation's overall earnings declined by 8 percent, Missile Systems Division sales rose more than 300 percent, to $24.1
million. Lockheed now decided to move to still larger quarters. In 1956,
it began construction on a $20 million facility to be spread between Sun-
nyvale, California, near the NACA's Ames Aeronautical Laboratory, and
nearby Palo Alto, the home of Stanford University, with its world-class
engineering faculty and yearly crop of well-trained engineering and science graduates. Meanwhile, the number of Missile Division contracts
multiplied. By the end of 1954, it was six; by the end of 1955, twelve. Its
concentration of scientific talent won Lockheed the Air Force's technically challenging WS-117L reconnaissance satellite in October 1956. In
February 1957, the firm got the contract for the Navy Polaris ballistic
missile.[24]

Nor were other companies, whether in aircraft, electronics, chemicals, or rocket motors, inactive. Olin Mathieson Chemical Company
bought a 50 percent share in Reaction Motors in 1953. In 1955, together
with the decade-old Marquardt Aircraft, it created a joint research organization called OMAR to develop products like rocket fuels and heat-
resistant structural materials.[25] Grand Central Aircraft Company founded
a rocket division in 1952, building it around a former chief of the solid
propellant division at the Jet Propulsion Laboratory and six younger JPL
rocket engineers.[26] Bell Aircraft Corporation, which in the 1940s had contracted out the rockets for its experimental planes to Reaction Motors,
built its own rocket research facilities and plant.[27] General Electric established a Missile & Space Vehicle Department with 400 employees in 1955.
Its first contract was an Air Force project for a nose cone that would allow
an atomic warhead fitted to a missile to re-enter the earth's atmosphere

intact.[28] Douglas Aircraft Company, contractor for the Thor intermediate range ballistic missile, by late 1957 decided the time had come to organize a separate Missile and Space Division.[29] There was, in short, a grand expansion of the new space industry from the start of the Korean War to the time when Sputnik was launched.

The very process of bringing into the companies the additional skills that were needed generated its own kind of tension. In December 1955, a dispute broke out at the research laboratory Lockheed created for its Missiles Systems Division over whether scientists or project managers should have control of advanced development. Fifteen top scientists and laboratory heads resigned as a result. The researchers wanted to direct those projects for which "the skill and technical knowledge [was] beyond the state of the art," but "'Top management said No,'" explained one of those who resigned. It wished the projects instead to be managed by groups outside the laboratory, groups that would call on the laboratory for research assistance. "Essentially, this would make the research laboratory a kind of service organization to be called in for help when problems arise."[30] Research scientists and production engineers did not always coexist happily.

THE MILITARY handled its space systems contracts with industry in a variety of ways. In 1952 the Air Force introduced the "weapons system" contract. A single prime contractor would be chosen and given wide responsibility for each system, including design, development, procurement of subsystems, and integrating systems into the final missile or plane. The weapons system concept was a recognition that the postwar generations of Air Force hardware required more attention to systems integration. It was no longer satisfactory to hire an aircraft company to build an airframe and then have them stuff in subsystems—radio, radar, guidance systems, engines, and so on—that were furnished by different specialized Air Force offices. Interactions among subsystems had become too important. For example, propellants had to be mated to structural frame elements that could withstand corrosive effects, the extreme cold of liquid oxygen, or the extreme heat of combustion. Engineers had to decide whether it was sounder to keep missile weights down by trimming the structure, shaving the electronics, or finding a propellant that could deliver the same thrust at a lesser total weight.[31]

Weapons system procurement was popular with the large companies who stood to be prime contractors. Second-tier producers, those who would furnish the subsystems, were less enthusiastic. They feared the primes would take over their business, because under this system the primes would have access to the proprietary know-how of the second-tier firms and thus would be able to pull jobs away from their subcontractors and perform them in-house.[32]

The Air Force was sensitive to this problem. In particular, it did not want the aircraft companies that served as primes to go out and acquire the expertise to bring jobs in-house. It feared that if the aircraft industry were to hire massive numbers of electronics scientists and other experts, the base of knowledge in other industries would be depleted. The airframe companies would have only second-rate competence in the new fields, at least during the period in which they were building the programs, and the result for the military would be higher costs and delays. Instead, in the words of one anonymous "top Defense Department executive," an airframe company serving as prime contractor should "bring in expert contractors in the fields in which he is not expert."[33] North American Aviation President Atwood recalled that Air Force Secretary Harold E. Talbott was dead set against North American's creation of four specialized divisions—Rocketdyne, Atomic International, the Missiles System Division, and Autonetics—out of its Missile and Control Equipment group. North American's leaders found it prudent to delay its reorganization until Talbott left office; within a week after he left they had effected the change.[34] As the anonymous Defense Department executive quoted previously put it, "There have been some people, some contractors, who imagined that if they didn't know a particular game they could build up the competence to do it in a fairly short time . . . [but in the United States] we are industrially organized, different industries have different special competencies, and I think we in the military ought to try to pull into these projects the best special competencies that are needed to do the job."[35]

In inaugurating its Atlas ICBM program in 1954, the Air Force and its scientific advisors moved away from the practice of putting prime contractors in charge. One example is its stance toward Convair, the company contracted to manufacture the Atlas missile. Convair engineers had been working on the Atlas since 1946 and had made a number of important and original contributions. Nevertheless, the Air Force doubted that Convair had the expertise needed to oversee the whole missile. More generally, it believed that the aircraft industry as a whole lacked the scientific and technical talent to take on that responsibility for missiles.[36]

How then, was the Air Force to effect the integration of the Atlas? It did not choose to build an in-house systems engineering capability of its own. Instead, it reached out to two scientists, Simon Ramo and Dean Wooldridge, who had headed Hughes Aircraft Company's Research and Development Laboratories, but had left Hughes in 1953 to start a company of their own. The Air Force arranged that the Ramo-Wooldridge Corporation would handle systems engineering and technical direction for the program. Convair became one of a number of associate contractors, the one charged with designing and building the airframe and fitting the other systems into it. Other associates were General Electric and A.C. Spark Plug for the guidance systems, North American for the engines, General Electric for the nose cone housing the warhead, and Burroughs for the computer.[37]

The Air Force decision to make use of Ramo-Wooldridge was controversial. Members of the House Committee on Post Office and Civil Service worried about the unusually high salaries and contract fees that Ramo-Wooldridge had negotiated. In addition, Congress asked why the Air Force, which had come into being in 1947 with none of the internal technical resources of the Army and Navy, had not elected to build up its in-house capabilities for this project. Granted that employing firms like Ramo-Wooldridge was a way of enlarging the nation's industrial base, one also had to reckon with the fact that private industry, precisely because it was private, could always choose not to take on one or another government project. Shouldn't a program of such importance and duration be directed by government scientists?[38]

The aircraft industry also protested. In its eyes, by setting Ramo-Wooldridge in this position and funding it so munificently, the government was using taxpayer dollars to create a rival company.[39] (In point of fact, Ramo-Wooldridge did become a powerful aerospace company. In 1958 it merged with Thompson Products, Inc., the company that had provided financing for its organization and expansion, to become the firm of TRW.) Nor did many of the companies like the idea of having systems engineering taken from them. The more they could cultivate this as an in-house skill, the better their position. In an age of complex systems, companies strove to be "systems houses."

The Navy also used the weapons system contract, though to a lesser degree than the Air Force. Of particular interest here is Project Vanguard, because its team was destined to become part of NASA. Project Vanguard became a microcosm that encapsulated every imaginable tension between a government agency and its contractor.[40] The Naval Research Laboratory and the Glenn L. Martin Company fought over Martin's demand that it be given full responsibility and NRL's contrary insistence that the laboratory retain control. Each harbored doubts about the other's competence. Martin saw the NRL as a nest of researchers eager to waste time analyzing the reasons for every failure or success, always promoting the "better" at the expense of the "good enough." NRL thought Martin engineers did not grasp how much they were dealing with unknowns, nor the importance of reliability and launch-worker safety. Martin wanted complete freedom within the specifications agreed upon to design and build the vehicle as it saw fit. NRL insisted that company-instituted changes be sanctioned by the laboratory, and that Martin abide by NRL-instituted changes, even if the company thought them unnecessary or wrong-headed.

Martin Company personnel thought they were oversupervised. NRL's Project Vanguard staff numbered 180, of whom 50 to 60 were engineers or scientists. The team of design engineers to be overseen at Martin only numbered three hundred. Martin engineers complained about "hit-and-run engineers" from the laboratory. NRL thought the supervision necessary and useful. It demanded, as well, the right of surveillance over Martin's subcontractors, so that the subsystems might also be kept to the highest standards. Both sides were made edgy by the knowledge that they had little more than two years to reach a successful launch and that the race, moreover, was not just against time but against the Soviet Union, which, in late 1956, announced its own plans to launch a satellite.

Vanguard historians Constance Green and Milton Lomask point out that the Martin Company took on the Vanguard project for prestige, as it did not expect that providing the government with a dozen or so Vanguard rockets would make money. To this we might add that Martin was also working under Ramo-Wooldridge on the Titan missile. The company was angling to be allowed to do its own systems engineering on future Titans, thus remaking itself into a space and missile company. In the matter of Vanguard, then, prestige rather than profit was a sound business goal. If Martin could demonstrate its proficiency in systems engineering and management on the Vanguard, the company stood to benefit when it counted—on Titan and other missile contracts.

In its Army Ordnance Missile Command (in 1956 reorganized into the Army Ballistic Missile Agency) and the Jet Propulsion Laboratory, the Army had a depth of expertise and facilities, and so did not rely to as great an extent on industrial contractors as did the Air Force and Navy. For the Redstone liquid-propellant missile, the Army Ordnance Missile Command at Huntsville built the first sixteen in its own facilities, using them as test vehicles. Only then did it turn the job over to a contractor, which built the last twenty test models, and all the operating missiles, under tight Huntsville supervision.[41]

The Jet Propulsion Laboratory not only took on the design, development, and prototype construction for the Corporal, its first major missile project, but also involved itself in the details of the production process at the contractor's factory. When the finished missiles were delivered, JPL crews dismantled them, inspected them, made modifications to the components, and then reassembled the rockets.[42]

The Army did not maintain this degree of involvement in all its missile projects: it did not have the manpower. Even Huntsville occasionally farmed out designs: for example, it gave the Pershing solid-propellant missile to the Glenn L. Martin Company. But as a general matter, the Army Ballistic Missile Agency strongly preferred an arsenal system. In-house work, the Army maintained, gave its staff the knowledge needed to supervise contractors effectively, and the capability to come to their aid if problems developed.[43]

It goes without saying that many companies liked this "arsenal system" of making missiles as little as they liked having an outside firm like Ramo-Wooldridge assume technical direction. It was an insecure arrangement for the contractor because the contracting agency had complete technical know-how and owned the engineering drawings, making it far easier for the work to be transferred to another company. It was an arrangement that threatened to deskill a contractor's engineering force. It was, finally, ignominious, and pride can be intimately connected with good business strategy.[44]

Industry's position was made more complex in that the Army and Air Force were locked in a fight over the Jupiter versus the Thor intermediate range ballistic missiles. Each branch attempted to enlist on its side the industrial firms most closely connected to it. For the Air Force, this meant the aircraft companies and their trade group, the Aircraft Industries Association. The result was that the Army's antagonism to the Air Force overflowed into an enmity between the Army and the aircraft industry.[45]

It is significant in this regard that neither the Corporal nor the Redstone and Jupiter contractors came from the aircraft industry. The Corporal's contractor was Firestone Tire and Rubber, one of the nation's leading manufacturers and retailers of tires and a firm that was trying hard to get into missile work when it received the 1951 contract.

Redstone and Jupiter were manufactured by Chrysler, an automobile company with a long history of Army contracts. Historically, the automobile industry had many connections with the aircraft industry. It was the matrix from which many early airplane inventors emerged and it had been called upon to help in the production of military planes in the world wars. Its executives had the habit of organizing or buying into an aircraft company from time to time. Nevertheless the aircraft sector had always viewed automobile companies with unease, as potential rivals and as populated by managers who did not truly understand the complexities of air and space transportation hardware. In Army ranks and in the aircraft

community, the arsenal system, the Jupiter missile, and the missile's manufacturer were at times seen as a single unified complex standing in opposition to another indissoluble entity: the Thor, the Air Force practice of
reliance on contractor expertise, and the aircraft industry.[46]

For the aircraft industry, the stakes were high. It looked as though aircraft were being replaced by missiles as the systems the military would be
buying. If the companies wanted to make money, they needed missile contracts. Yet there was no a priori reason why the military should use them
as prime contractors for its missiles rather than chemical companies like
Thiokol or electronics companies like General Electric or Western Electric or automobile companies like Chrysler. Thus the aircraft companies
fought hard to claim space as their turf. In 1955, Lockheed president
Robert E. Gross' first major address as president of the Institute of Aeronautical Sciences was full of such territorial claims. He warned that outside firms were invading the design and production of aerial weapons systems—under which heading he placed both military aircraft and pilotless
aerial weapons like missiles. "Technical advances and new concepts have
brought other companies into the picture, he commented. . . . 'We do not
necessarily have within the geographic confines of the earlier airframe
companies the people with all the skills now needed.' . . . Airframe companies . . . must stand firm against this outside invasion, Gross insisted. . . .
Development of aerial weapons belongs with the airframe companies."[47]

Van Dyke and others have pointed out the polemical nature of the
term "aerospace" in the 1950s. It was used by the Air Force as an assertion of its title to missions in space as a natural extension of its role in the
air. It was equally a claim by the aircraft industry that it should be the primary sector to make the space systems.[48]

The NACA Heritage

The National Advisory Committee for Aeronautics—NACA—was created in 1915 to supervise and coordinate U.S. aeronautical research. During the 1920s, it evolved into a research agency, one intimately tied to the
airframe and engine industries. The data its engineers generated were
warmly received and widely used by industry engineers, data on matters
like the aerodynamic properties of different airfoils. But the NACA did
not contract with industry, nor did it ordinarily design specific planes. Instead it did "generic" or "fundamental" research, applicable across a range
of designs.

This NACA policy of concentrating on generic research had been
codified at the end of World War II. The policy was a product of negoti-

ations among the NACA, the aircraft industry, and the military, and it reflected the relative strength of these three players. Aircraft had entered the war ranking forty-first among U.S. industries; by 1944, it was the largest sector of U.S. manufacturing.[49] With its new prominence came the power to further its own agenda, and one of its aims was to keep as its particular province the design and development of airframes.[50]

What industry did want the NACA to do was to build research facilities like elaborate wind tunnels that individual companies were unable or unwilling to finance. It wanted the NACA research engineers to run its models in the tunnels, and to troubleshoot its prototypes. And it wanted the NACA to do generic research and to make the results of that research known promptly to the companies. The NACA's engineers enjoyed fundamental research, but like any engineers, they were also passionately interested in design. Nor were the NACA's leaders sure that the design of truly advanced planes and engines could be left to industry. They feared that private companies would be content to confine themselves to being a mere one step ahead of competitors. Experience underlay this apprehension. In the early stages of World War II, the established aircraft engine companies had declined to take up jet engine research or even the liquid-cooled piston engine out of conservatism and a fear it would jeopardize profits.[51]

The NACA did not have a strong ally in the third player in the negotiations, the military. For one thing the military harbored some distrust of the NACA as it went into the postwar era. The NACA had been almost as slow as the engine companies to anticipate the importance of jet engines for military aircraft in the 1930s, with the consequence that the United States lagged behind Germany and Britain in jet airplane development. In addition, the military was beginning to strengthen its own facilities for doing aeronautical research, and its R&D organizations were coming into jurisdictional conflicts with the NACA laboratories.[52]

In 1946, after two years of discussions, the NACA formulated a policy statement reflecting the tug-of-war that had gone on. It affirmed that its "principal objective" was to provide "[f]undamental research in the aeronautical sciences. . . . Application of research results in the design and development of improved aircraft and equipment, both civil and military, is the function of industry." To the military was left "[t]he evaluation of military aircraft and equipment developed by the industry, and the exploration of possible military applications of research results."[53]

Two experimental supersonic planes of the mid-1940s, the Army Air Forces' XS-1 and the Navy's D-558, illustrate the division of labor. The NACA worked with the Army and, more closely, with the Navy in specifying the general guidelines for these airplanes—particulars like whether to use rocket or turbojet propulsion and straight wings or back-swept wings. The translation of these overall specifications into designs was the province of the engineers of the Bell Aircraft Corporation for the XS-1 and the Douglas Aircraft Company for the D-558. Engineers at the NACA took charge of the instrument package, conceiving and executing the devices that would provide data on the behavior of aircraft at high speeds. Their instruments measured things like air pressure, aerodynamic and gravitational forces, acceleration, vibrations in the structure, and engine performance. The Army Air Forces and the Navy supervised the contracts, paid the bills, and had the final word.[54]

In the mid-1950s, the NACA broke with this tradition. Engineers at NACA's Langley Research Center in Virginia carried out extensive experimental and paper studies in 1953 and 1954 to define the features of an airplane that could fly at altitudes of up to 50 miles, and at speeds up to 4,500 miles per hour. The NACA had enough results by mid-1954 to sell its design of an experimental plane, called the X-15, to the Air Force and the Navy. This time NACA engineers had technical responsibility, with the power to pass on contractor designs, and the task of working with contractor engineers on every phase.[55] Again, however, it was the Air Force that ran the contract, and the Air Force and Navy that together paid the bills.

Parts of the NACA had been eager to add space to aeronautical work since the closing days of World War II. As the years went by, the agency increasingly engaged in space projects. In general, however, it found itself restricted to roles that followed the division of labor laid down for aeronautics. The NACA did generic research: an outstanding example was its success in devising a shape for missile warheads that would keep them from burning up when they reentered the atmosphere. And it did troubleshooting to clear up problems with specific missiles; for example, NACA aided the Army in eliminating fires and explosions in the Jupiter missile's engines.

Overall, industry had a particular image of NACA engineers. It saw them as researchers, people whose aim was the production of papers and books. In contrast, industry saw itself as product-oriented; as willing to steal ideas, take risks, extrapolate beyond the data, in order to get a product satisfactorily designed and promptly delivered.[56] It was the old dichotomy between research engineer and project engineer that underlay the fight at Lockheed's Missile Systems Division laboratory and the tension between the Martin Company and the Naval Research Laboratory. In the case of the NACA, it was not without irony, for the relegation of

NACA engineers to generic research after World War II had come precisely at the insistence of industry.

The Establishment of NASA

The Naval Research Laboratory's Project Vanguard team did not succeed in putting up the first artificial satellite of the International Geophysical Year. Instead, this race was won by the Soviet Union, which launched the Sputnik satellite on October 4, 1957. It was a turning point for the U.S. space community. Sputnik led to a proliferation of new or refurbished schemes for space projects, and an outpouring of suggestions for the organization of U.S. science and technology in general and the space program in particular. Government agencies began jockeying for major roles in the larger space program that was surely coming. Private companies began augmenting their space expertise and space facilities.

Among government agencies, the Air Force moved rapidly and boldly. The Secretary of the Air Force immediately set up a committee with members drawn from Air Force ranks, industry, and academia. By October 28th, the committee submitted a recommendation for a unified national space program under Air Force leadership. Meanwhile, the Air Force's Scientific Advisory Board had recommended that the Air Force field military reconnaissance satellites, meteorological satellites for the determination of weather conditions on battlefields, and communication satellites as soon as possible to enhance Washington's ability to command and control U.S. forces world-wide.[57]

The Scientific Advisory Board had also recommended that the Air Force plan toward eventually undertaking lunar expeditions with human crews. At the same time, the Air Force took immediate steps to implement robotic lunar explorations, sending out a call to industry for proposals. It also began to explore promoting one of its near-Earth space projects into a lunar project. This was Project Farside, begun in about 1956 under the auspices of the Air Force Office of Scientific Research. Its object had been to send a rocket 4,000 miles from the Earth's surface to measure magnetic and gravitational fields, cosmic radiation, and meteorites. Now Air Force agencies began to look into a possible Phase II to Project Farside that would send an instrumented packet by the moon.[58] The Army also wanted a major role in space, based on its research teams at the Army Ballistic Missile Agency and the Jet Propulsion Laboratory. It also had projects, among them a proposal called Project Red Socks that originated at the Jet Propulsion Laboratory, for a series of nine circumlunar flights, commencing in June 1958.[59]

On November 3, 1957, the Soviets sent up a second satellite, weighing more than 1,100 pounds and carrying a dog. It was now patent that they were planning to put a person into orbit. If the United States wished to regain its reputation as the world's technological leader, it looked as though it would have to race the USSR to be the first to send humans into space. The two Sputniks also underlined to the U.S. defense establishment the importance of space as a theater of military operations. Exactly what kind of operations was not clear. Ideas of all kinds were advanced. Should there be space-based bombs, or space-based anti-missile missiles? Was it necessary to take steps to deny all access to space to unfriendly nations? Or would it be sufficient to have control of space in times of war, similar to the control of the air and the seas that classic military doctrines called for? And in that case, was it indeed possible to create the capability to intercept and destroy harmful spacecraft?[60]

The Air Force was particularly insistent that it would be necessary to place military personnel in space. As early as the beginning of 1956, the Air Force Research and Development Command (ARDC) had advocated this. Specifically, it had sought a follow-up to the X-15 plane, and had recommended studies of a piloted space glider and a "ballistic" spacecraft. The latter was essentially the nose cone of an ICBM, converted from nuclear warhead to a capsule housing humans. The Air Force upper echelons had declined to fund these studies in 1956. Now Air Force research and development leaders reintroduced these ideas and many others as well, including manned satellites in geosynchronous orbits,[61] 22,300 miles from the Earth, and a lunar base staffed by the military for surveillance and bombardment.[62]

Shortly after Sputnik II, the Secretary of Defense took steps to set up within his office an agency with jurisdiction over space programs. His aim was to centralize the work, and to defuse the Air Force-Army rivalry. By February 1958, the office, the Advanced Research Projects Agency, had been formed. The Air Force opposed it. The Army embraced it as a way to protect a role for itself in space and to advance its proposals for larger rockets, heavy reconnaissance satellites, and expeditions with and without humans on board.[63]

WHILE THE military services were scrapping for control of the national space program, industry was proffering its own suggestions for projects. Some of these were ideas that companies had already been studying before the Sputniks went up, often in close connection to military projects. AVCO Research Laboratory, for example, was one of the Air Force Air Research and Development Command's nosecone contractors. When the

ARDC failed to get funding for contracts for a ballistic capsule based on nose cones and using ICBMs as boosters, AVCO, with ARDC encouragement, continued the project using company funds. AVCO accelerated this work after the Sputniks were launched and, in November 1957, the company submitted a new proposal to ARDC. This also was not funded, so AVCO went to the Convair Division of General Dynamics, maker of the Atlas ICBM, with the idea of formulating a joint proposal for an AVCO capsule with an Atlas booster.[64]

Convair Astronautics Division of General Dynamics had had a group working for several years to evaluate space projects, under the direction of Krafft A. Ehricke, former member of Nazi Germany's V-2 rocket team. At the Senate's "Inquiry into Satellite and Missile Programs," Ehricke was able to show the senators small-scale models that Convair had built of a manned satellite, a space station, and a vessel for landing parties on the moon, or flying them to Mars and Venus. Ehricke also took the occasion to advocate a new rocket stage, using liquid hydrogen as propellant, that could be added to the Atlas to give a booster capable of putting these spacecraft into orbit.[65]

Industry executives and top scientists were also called on by the administration, the military, and the Congress to serve as "elder statesmen" as the nation decided how to respond to the Sputnik launches. Unfailingly, they advocated reversing the cutbacks in weapons procurement that the Eisenhower administration had recently imposed. Robert E. Gross, President of Lockheed, portrayed the United States as in a crisis almost as grave as actual war. More dollars were needed for national security; aircraft as well as missiles needed to be funded. Palpable in his testimony was his sense of devotion to the public weal.[66] (The increased expenditure he and the others called for was bound to benefit the aerospace industry, but there is an alchemy, nigh universal, that operates to transmute private interest into public good.)

Industry leaders also brought to the table a slew of plans for organizing the U.S. space effort. These were men who were deeply familiar with federal military R & D. Some decried compartmentalization. Many seconded the opinion of the military services that the space program be lodged within the Department of Defense. Some, like Simon Ramo of TRW, supported proposals for a cabinet level Department of Science that would house space as well as other R&D operations.[67]

THROUGH THE winter of 1957-1958, military agencies continued to lobby hard for control of the space program. In February 1958, the administration and Congress gave the Advanced Research Projects Agency the authority to conduct for one year presidentially decreed civilian space projects as well as military space projects. But that was a temporary measure. By February 1958, presidential science advisor James R. Killian and President Eisenhower were convinced that the scientific and civil part of space ultimately had to be placed under a civilian agency.[68] For Eisenhower, who was chiefly interested in military uses of space, a civilian agency lofting civilian satellites would help establish an "open skies" principle vital for the future use of military reconnaissance satellites: the principle that satellites could fly unimpeded over any nation.

There were several candidates for the core of such a civilian agency: the NACA, the Atomic Energy Commission, the National Science Foundation, a proposed Department of Science, or a new space agency cut out of whole cloth. The NACA had some attractive features. It already existed. Its engineers by this time were as occupied with space and missile projects as with aeronautical ones, and they had acquired considerable skill in the field.[69] Finally, a civilian agency would need to know how to work in close cooperation with the military, and the NACA had a long history of working with the services. At the beginning of April, the president sent Congress a bill for a National Aeronautics and Space Agency (NASA) that would use NACA as its core.[70]

The administration bill gave NASA a structure different from the old NACA. It replaced a director appointed by the advisory committee with a director appointed by, and responsible to, the president. Like the NACA, the new NASA was to have a strong research program: "The Agency shall . . . develop a comprehensive program of research in the aeronautical and space sciences [and] . . . plan, direct, and conduct scientific studies and investigations of the problems of manned or unmanned flight . . . with a view to their practical solution." But NASA was also to "develop, test, launch, and operate aeronautical and space vehicles." In other words, NASA was free from the limitation to generic research that had confined the NACA, and could mount its own hardware projects. Finally, NASA was authorized by the bill "to enter into and perform such contracts . . . as may be necessary . . . with any person, firm, association, corporation, or educational institution." The old NACA had been restricted to a maximum of a half-million dollars per year of contracted work.[71]

Neither the proposal to allow NASA design and development projects nor that expanding its contracting powers went without criticism in the hearings that followed the bill's submission to Congress. On development projects, the NACA itself was divided. Many at headquarters and at NACA's Ames Research Center preferred that NASA inherit the NACA's culture of fundamental research, while a sizeable number of engineers at

Lewis and Langley Research Centers were eager to get their hands on projects.[72] Opponents of transforming the NACA into a development agency argued that the NACA engineers were temperamentally unsuited to development. A report issued by the House Select Committee on Astronautics and Space Exploration summarized some of the testimony. "N.A.C.A., as now constituted, is a research agency, with the traditions of a research agency. . . . its operating traditions have all been consultative, advisory, mediatory. . . . Without drastic, sweeping changes, it is no mean feat to inculcate a spirit of decision-making in an organization that has lived and thrived on a tradition of peaceful advice-giving."[73]

The most prescient objection to development was made by Walter R. Dornberger, formerly head of Hitler's V-2 proving grounds and now technical assistant to the president of the Bell Aircraft Corporation. Dornberger argued that a NASA that had projects of its own would inevitably slight those areas of space and aeronautical research that were not germane for its hardware. "A scientific research organization with the authority to develop its own projects will tend to back its own projects exclusively."[74]

Some military officers and members of Congress voiced the fear that too many agencies authorized to contract with the same defense firms would lead to competition and drive up prices. Whereas as recently as World War II, only the Army and Navy were letting contracts, now there would be six agencies: the three services (Army, Navy, Air Force), the Advanced Research Projects Agency (ARPA) that the Secretary of Defense had just created, the proposed office of the Deputy Defense Secretary for Research and Engineering, and NASA. Even worse than price competition for these critics was the possibility that military needs might go unmet when the many agencies drew on a single, limited pool of industrial resources. If a civilian agency must be formed—and many of these witnesses preferred that the space program be lodged in the military—then it might better contract through the services, as the NACA had done for the X-15. The Defense Department could ensure that top priority go to those projects that were most vital to national security.[75]

Industry naturally saw this proliferation of contracting agencies in a different light. Here we must distinguish the statements of business leaders—many of them supported military control of the space program—from their business strategies. Were there too much work, the worst prospect aerospace companies would face was that of having to forego entering a competition for a project they would have liked to have.

Indeed, the creation of NASA as a contracting agency had some distinct advantages. It opened a way into space for companies that had not yet participated in missile and rocket development on any substantial level. Every contracting agency had a coterie of preferred contractors, companies that its program managers had come to know and trust, that had achieved a proficiency in the relevant technology through having received and capably executed the earliest contracts in the field. Agency familiarity with firms inside the circle created a barrier to entry for those on the outside; conversely, the establishment of a new contracting entity would level the field.

Hughes Aircraft Company is an example. At the time that Simon Ramo and Dean Wooldridge left Hughes in 1953 to go into ballistic missile management, and perhaps in part as a reaction against that defection, Hughes' leaders had concluded that ballistic missiles were not technically promising. Hughes had missed the first call to space, and when NASA formed, it resolved to correct its error. "We then started going after NASA projects. NASA was a new organization, and we had as good a footing as anyone with it. It was a field that was not yet occupied. . . . In contrast, we were outsiders in Air Force space programs."[76]

A second advantage that NASA could confer devolved from the open nature of most of its projects. This would be space work that one could boast about, that one could use externally to build up the company image and internally to inspire the work force. Military space projects, in contrast, were often highly classified; where they were not, it was still difficult to make good public relations out of the kill-power of a missile.

The number of spacecraft that NASA would order from industry would undoubtedly be small. Defense leaders did not expect that NASA and the military combined would need more than a few thousand space vehicles in all over the next twenty years.[77] In addition, firms would have to fund expensive new research and production facilities, and work to much more stringent standards of production. Twelve years earlier most companies had turned up their noses at the Army Air Forces-Navy-NACA experimental airplanes. Only a handful of planes were to be ordered and airframe companies were used to making their profits on large production runs.[78] But times had changed. The Atlas, Thor, Titan, and Jupiter missiles that defense firms were contracted to build numbered only in the tens and hundreds. Building them also required new facilities, and imposed a degree of cleanliness in production, of temperature control, and of precision in machining that far outstripped World War II aircraft production. The amount of research and development that a defense firm was obligated to put into a product had gone up steadily. The ratio of engineer and scientist to production worker had soared since the start of the decade.[79] Thus, the disadvantages that would be incurred by working for NASA represented an intensification of trends already underway, rather than a radical departure from practice.

Working for NASA would give companies the chance to learn technologies, develop skills, and install production tooling that they could use for other projects. In the first instance, this meant military projects. In the spring of 1958, there was substantial congruence between the kinds of projects a NASA was likely to pursue and those the services wanted. Both could be expected to mount manned and unmanned satellites, lunar, and even planetary missions. Beyond its transferability to military work there was the expectation that some of the technology would spill over into commercial products.[80] Finally, there was another reason why the aerospace industry was interested in working for the civilian space agency that was aborning. This was the enthusiasm and interest in space that Sputnik had kindled. To begin with this took the form of a groundswell of excitement among bench-engineers. But it also reached up into management. The process of replacing the aviation enthusiasts who had founded the industry—the Glenn Martins, William Boeings, and Donald Douglases—with financiers, lawyers, and managers in the top echelons of aerospace was already well underway.[81] But in those levels there were still men who were basically engineers, and they were just as susceptible to technological fervor as the men and women in their development divisions.

In July 1958, after debate, emendations, and deals, the act establishing the civilian space agency was signed into law. In its passage through Congress, changes had been made in the structure of the organization, and there was a slight change in name, though not initials, to National Aeronautics and Space Administration, but its license to run its own projects and to contract much of its work to industry had been preserved.[82]

NASA and Industry

How the space program should be divided between government and industry was a disputed matter within NASA. Each of the organizations entering the new agency brought its own culture and perspective. The Jet Propulsion Laboratory, with close to 2,400 employees, joined NASA at the end of 1958 with visions of becoming the agency's premier research arm. It had a strong in-house design and development capability and it was open about its desire to continue to immerse itself in R&D and to minimize its contract-monitoring duties.[83]

About 200 staff members from the Naval Research Laboratory, among them the Project Vanguard team, were being transferred to the NASA Goddard Space Flight Center then being built just outside Washington D.C., in Maryland. The Naval Research Laboratory was also an institution with a tradition of in-house research. Project Vanguard's team had worked with an outside contractor, the Glenn L. Martin Company, but it had forged a policy of close supervision and control over its contractor. As we have seen the NACA, with its staff of nearly 8,000, had little experience in supervising industry contracts.

In 1959 and 1960, NASA negotiated to bring into its organization the almost 4,700-strong Developments Operation Division of the Army Ballistic Missile Agency in Huntsville. This team, headed by Werner von Braun, had successfully engineered the Redstone and then the Jupiter missile and had just begun designing a super-booster for the Department of Defense, the Saturn. It would bring NASA expertise in large launch vehicles, but it would also bring a group identified with the arsenal system of in-house design and systems integration.

The new NASA administrator, T. Keith Glennan, was not a member of any of these organizations, but an outsider. Glennan, in fact, had started out in the movie business. He became a manager of science and technology during World War II, when he served as a director of the Navy Underwater Sound Laboratory. He did a two-year tour of duty as a commissioner of the Atomic Energy Commission from 1950 to 1952, but his chief postwar position had been as president of Case Institute of Technology in Cleveland, Ohio.[84]

Glennan was well aware of the attitudes of the organization over which he was now presiding. And he took seriously his responsibility as a steward of the in-house capabilities he had inherited.[85] At the same time, Glennan faced an industry that was closely watching NASA's contracting practices and was apprehensive that the growing space program would be executed within government laboratories, rather than translated into business for itself.[86]

Glennan favored contracting out as much as possible. An Eisenhower Republican, he believed in keeping the federal government small. Contracting would educate industry in space technology, and thereby give government an industrial sector it could turn to in the future for its space needs. In addition, commercial use of space for communications seemed to be coming faster than anyone had expected. Here, too, contracting out space work would help prepare industry to succeed in the new business area.[87] "It is natural," Glennan wrote in his diary in April 1960, "that our scientists and engineers want to keep to themselves all of the interesting and creative problems, while farming out to industry only the repetitive and straight production items." But he added, "[t]his does not make sense to me."[88]

From October 1958, when he took up NASA's reins, Glennan had employed the consulting firm of McKinsey and Company to help him structure the agency's management. In 1960 he ordered a pair of reports from McKinsey; one was on the proper role of NASA vis-à-vis its contractors. Glennan worked closely with the authors and the project became a way of crystallizing his intuitions and sounding out the reasoning behind them. The reports constitute a window into his thinking.[89]

The report on NASA-industry relations recognized two central dilemmas: how to make good use of NASA's inherited capabilities, while still giving private enterprise a leading role, and how to control industry sufficiently to get equipment with exceptional reliability, yet not control it so much as to stifle its creativity.[90] The answer was that NASA should retain for the time being preliminary and conceptual design and systems integration. Simultaneously, NASA should contract out as much as possible, using the contracts to educate industry, and gradually transferring as much of the space program as practical to this educated industry. Thus, study contracts for preliminary and conceptual designs should also be awarded to industry "to round out industry's capabilities." The authors affirmed: "Unless industrial contractors are encouraged to round out their capabilities, NASA will find it necessary to expand its in-house capabilities." Expanding in-house laboratories, and thereby government payrolls, was something Glennan definitely opposed. "[A practice of maximum contracting] will mean that NASA will depend to a much greater degree, in the future, on industry to schedule, coordinate, and physically integrate complete space vehicle subsystems."[91]

The report was circulated to the centers, from which came a number of objections. Center leaders pointed out that the industrial policy it embodied might allow a few large firms, those with systems integration capability, to dominate the market. They feared that farming out too much of any program would threaten to erase the collective memory necessary to preserve the continuity of space work because, in contrast to civil servants, industrial teams disbanded once projects were over and industrial engineers frequently moved from one firm to another. They worried too that, given too much authority, industry firms might do shoddy work.[92]

WHILE THESE internal discussions went on, NASA's first high-profile venture, Project Mercury, was leading industry observers to some preliminary opinions of their own regarding NASA's relations with its contractors. The project, to get a human into space before the Soviets could, had been assigned to NASA by President Eisenhower in August 1958. The decision had been made at the top level, because the assignment was contentious. Early on, all three services had vied for the project. The Air Force had worked and reworked a scheme for a ballistic satellite to be launched by an Atlas missile, which it called Man-in-Space-Soonest. The Army's Huntsville arsenal had promoted a stunt it called Project Adam, which it believed could be carried off as early as 1959. Project Adam involved using a Redstone missile to send a manned capsule on a parabolic trajectory that would take it to a height of 150 miles, before it fell back into the Atlantic Ocean. The Navy had suggested a ballistic orbiter, MER I, that would convert into a glider as it reentered the atmosphere. When the Advanced Research Projects Agency was formed in February 1958, Man-in-Space-Soonest had been turned over to it. But no amount of interagency negotiation had persuaded the military to give the human space flight project to NASA, and the controversy had been kicked up to the White House.[93]

Just as in industry and the military, the NACA's engineers had done some study of manned satellites before October 1957. And like their counterparts in industry and the military, NACA personnel had been galvanized by the Sputnik launches to accelerate that work. By the time of a March 1958 NACA conference on high-speed aerodynamics, two proposals for orbital flight had crystallized, one in California at the Ames facility and the other at Langley in Virginia. Ames engineers favored a satellite shaped so as to receive some measure of lifting force in its motion through the atmosphere.

The Langley proposal was for a ballistic satellite. It came originally from a group of engineers, headed by Maxime A. Faget, from Langley's semi-autonomous Pilotless Aircraft Research Division (PARD). PARD engineers had been involved in a greater proportion of space work than any other part of the NACA. They designed and built rockets and nose cones, studied heat transfer problems like reentry heating, and worked at guidance and control. In late 1957 and early 1958, they provided research support to the Air Force Man-in-Space-Soonest project. Faget and his coworkers believed the ballistic capsule was hands-down the simplest, and most reliable system, and the one that could be put into orbit the fastest.[94]

The design competition was affected by differing conceptions of how NASA should be structured. The Ames group did not want to run a project and hoped that NASA would retain the old NACA role as research institution. The people at Langley did want a project. Unlike most of Langley's staff, who did generic research, PARD was used to projects. By May 1958, they had settled on a shape for the capsule and invented a couch to support astronauts against the forces they would experience during the flight. By the time of Eisenhower's decision in August, Langley was already testing the parachutes that would slow the capsule's descent through the atmosphere and was experimenting with a water landing.[95] Through September, the last month of the existence of the NACA, and October, the first month of NASA, the team continued to refine its ideas and draw up specifications for the capsule. Thus the preliminary and conceptual design for the Mercury capsule was done in-house and was ready when NASA originated.

Shortly after NASA was formed, the Space Task Group was created to handle the Mercury project. A former director of PARD, Robert R. Gilruth, was chosen to head it, and it was made a quasi-independent field unit, with temporary headquarters at Langley.[96] In mid-November, NASA issued a request for bids from industry. The companies found that it was not only an in-house design they were being given but one whose salient feature was its specificity. "Faget and associates had described in remarkable detail their expectations of what the capsule and some 15 subsystems should be like."[97] As one NASA spokesperson said, "We wanted to nail down precisely what we want."[98]

Why? There was a sense of urgency that the job be done on time so as to win the race against the Soviet Union, and be done right so that astronauts not be endangered. Another reason was that the Langley engineers mistrusted industry's ability to design something as novel as a spacecraft. Although the aerospace companies had built missiles, they had not yet built space vehicles for human flight.

Distrust, indeed, was the order of the day in the 1950s, when everyone was scrambling to master the science and engineering of space flight. It was the Air Force's mistrust of the airframe industry that led it to engage Ramo-Wooldridge. Project Vanguard was marked throughout by Martin and NRL engineers' suspicions of each others' competence. Langley judged industry to be "immature" in matters of spacecraft, while industry and the military were convinced they knew more about space flight than NASA did.[99]

Two words have acquired specialized meanings in conversations about contracts in the aerospace community. "Responsive" describes a company or a proposal that takes the contracting agency's design to heart and uses it as a basis. "Arrogant" is used for the company that thinks its technical ideas are better. The amount of arrogance an agency will tolerate is something an aerospace company wants to know, and the place a company will seek on the arrogant-to-responsive spectrum when it submits its proposals depends upon that as well as on the firm's traditions, experience, and situation. But arrogance in proposals is also one of the channels by which creative ideas flow from industry to government. The extreme specificity of the Mercury request for proposals suggested to industry observers that NASA could grow into an agency wedded to its own design ideas. Nor could NASA, with its formidable in-house expertise, be expected to use the weapons system type of procurement the Air Force had pioneered. NASA seemed to be emerging as a body that would keep design and technical direction for itself.

A Tale of Two Companies

RESIDENT JOHN F. KENNEDY'S DECISION, IN MAY 1961, to land Americans on the moon within the decade of the 1960s made NASA into a bigger game in space. Whereas in fiscal year 1960 the Department of Defense spent about five dollars on space for every four that NASA spent, by fiscal year 1964, NASA was outspending the military in space by three to one.[1] Winning NASA contracts now became more than a matter of prestige and public image, or even of keeping on the frontiers of technology. It was big bucks.

As part of the race to demonstrate U.S. superiority over the Soviets, President Kennedy also emphasized creation of an American-built satellite system for world-wide communication, and additional monies were made available to NASA for R&D in that arena as well. Again, industry was directly affected, as NASA's programs helped U.S. firms enter the field of communications satellite manufacturing.

Two companies that benefited from the Kennedy space race were Hughes Aircraft Company and North American Aviation. Hughes went from upstart to the world's preeminent manufacturer of communication satellites. North American's rocky road took it through conflict and crisis, but eventually led to the position of NASA's leading contractor.

Hughes and the Syncom Satellite

The Soviet Sputniks stimulated a new look at communications satellites. It was clear that they were going to have both military and commercial uses. For the military, they would facilitate contact with far-flung craft: for example, they would help solve the old problem of how to maintain reliable contact between the ground and the fleet of nuclear bombers the Strategic Air Command kept aloft. For the commercial world, they would be valuable for telephony between Europe and America, augmenting the transatlantic telephone cables then being laid, for television and radio broadcasting, especially to developing nations, and for facsimile and tele-type transmission.

There were questions to be resolved before a system could be fielded, and many of them went to the heart of the relation between U.S. industry and the federal government. How much of the research for the commercial communications satellites would be financed, directed, or done by government, and how much by the private sector? Would a private industry arise to launch the satellites or would they be launched by government agencies? Would industry or government own and operate the systems?[2] Under these issues, which were argued publically and passionately out of deep convictions and as part of the rhetoric by which government agencies and firms alike advanced their own interests, lay other, less articulated questions. What private firms would enter into the manufacture and the operation of commercial satellites (comsats)? What strategies would they use to gain market share? How would government policies and actions affect the market positions of private companies? How would these policies and actions affect the technology that was chosen? The early years of U.S. communications satellites became a forum in which some classic problems in the division of economic activity between the public and private sectors were debated.

The federal government and industry acted soon after the first Soviet satellites went up. In the spring of 1958 the Advanced Research Projects Agency (ARPA) that President Eisenhower had just established within the Defense Department scoured the country for satellite companies and satellite ideas. From RCA (the Radio Corporation of America), ARPA adopted the suggestion of a simple communications experiment called SCORE (Signal Communications by Orbiting Relay Equipment). SCORE had a tape recorder at its heart so that messages could be beamed to it by one ground station and played back on the command of others. SCORE was placed in an orbit 900 miles high in December 1958 and it broadcast a tape by Eisenhower and transmitted a number of other messages in twelve days of active life.

Even before the SCORE project started, the Department of Defense had been working on a more sophisticated version called Courier, which was to orbit at about 750 miles. Courier would be launched in October 1960.[3] ARPA, meanwhile, started work on a number of more difficult, but more useful satellites. Among them was one called Decree, which was to be positioned in geosynchronous orbit, 22,300 miles high, and would be able to receive messages and rebroadcast them simultaneously.

Leading electronics firms got the contracts for these satellites, contracts that permitted them to advance their capabilities in the new field. Philco Corporation, which had the Courier contract, was best known to the public as a manufacturer of radios, televisions, and household appliances. Government electronics represented one-fourth of Philco's $350 million revenue in 1958, and the products it sold the government included research and development, microwave communications equipment, torpedoes, and missiles. RCA, which made the SCORE satellite, was founded after World War I to hold a pool of U.S. and foreign radio patents and had not gone into manufacturing until the 1930s. By the late 1950s, however, its sales, at nearly $1.4 billion a year, encompassed TV and radio broadcasting, ship-to-shore and international radio communications, and a wide range of consumer, industrial, and military electronics systems.[4]

The Bendix Corporation, which with General Electric held the Decree contracts, was a smaller firm, with only half RCA's sales. It had begun life between the World Wars as the Bendix Aviation Corporation and by the time of the Sputnik launches was selling the government missiles, electronic components, and planes, with 70 percent of its business in defense. General Electric, for its part, was the daddy of U.S. electrical companies, registering over four billion dollars a year in sales, and fielding products that ranged from light bulbs to complete electric power plants. General Electric was one of the first firms in rocketry and had early on established a special division for missile and space work.[5]

There was one company, however, that did not need government funds to help it into the comsat business. This was American Telephone and Telegraph (AT&T). With over seven billion dollars a year in revenues and eighteen billion dollars in capital, AT&T was the world's richest firm.[6] AT&T held more than 80 percent of the U.S. local telephone market and dominated the nation's long distance and international telephone market.

AT&T's research arm, Bell Telephone Laboratories (BTL), had been interested in communications satellites for some years. In early 1958, shortly after the Sputnik launches ushered in the satellite age, Bell Laboratories went to the NACA and the Jet Propulsion Laboratory and proposed a cooperative effort to orbit an experimental passive communications satellite. Scientists at the NACA's Langley laboratory had already been preparing a hundred-foot sphere of aluminized plastic that would ride about a thousand miles above the earth and measure the drag of the atmosphere. Bell Laboratories' leaders suggested this metallized balloon be used instead to reflect radio waves from one earth station to another. The NACA and the JPL liked the idea, and the project started, under the name Echo. When NASA opened for business in October 1958, it took over the NACA's part in Project Echo.[7] By 1959, Bell Labs scientists began paper studies of a second, more advanced, active system (Telstar), which would consist of tens of satellites, flying at an altitude of several thousand miles and retransmitting the waves they received with onboard repeaters.[8]

Across the country from the New Jersey facilities of the Bell Telephone Laboratories, the Los Angeles-based Hughes Aircraft Company, headquartered in Culver City, also embarked on comsat research. Hughes' major product at the time was electronics systems for Air Force fighter planes. The market for fighters, however, was threatened in 1958 and 1959, as missiles began to displace bombers for the delivery of explosives. Fighter planes were designed to intercept bombers and had no efficacy against missiles. In late 1958, the head of Hughes' Advanced Development Laboratory assigned a small team of engineers, under the leadership of Harold A. Rosen, the task of looking for new products. Team member Donald Williams proposed as one possibility geosynchronous navigation satellites. Rosen thought geosynchronous communications satellites were still more promising. At that time it was assumed that geosynchronous satellites would be heavy and complex spacecraft, but Rosen and Williams had a number of ideas they thought would make their satellite "bird" simple, lightweight, and capable of rapid development. In the fall of 1959, they had worked out these ideas sufficiently to take a proposal to upper management.[9]

By this time the government had canceled North American Aviation's F-108 fighter, for which Hughes had held the electronics contract. With it, half of Hughes' business disappeared: new products were more important than ever. Hughes management appointed a task force to evaluate Rosen and Williams' proposal and this group reported back with an enthusiastic endorsement. It recommended that the Rosen-Williams geosynchronous satellite be developed for commercial use with private capital. The catch was that, unlike AT&T with its deep pockets, Hughes did not have that kind of capital. Hughes now began to talk with other organizations, both private companies like General Telephone & Electronics, the nation's second-largest telephone company at one-twelfth the size of AT&T, and government bodies, like the Department of Defense and NASA, about a jointly funded project. Meanwhile, Rosen, Williams, and their colleagues kept refining their concept and got to work on a demonstration model. NASA was happy to give Hughes advice on matters like appropriate launch vehicles and the environment the satellite would encounter in space. But it was unwilling to partner Hughes in an experimental program. For one thing, NASA did not have the money. For another, it had an agreement with the Department of Defense to divide up communications satellite work. This agreement called for NASA to restrict itself to passive satellites like Echo, leaving the active ones to its sister agency. Hughes' was an active satellite.

By 1960, more and more companies were taking a look at the communications satellite business. Lockheed was considering manufacturing comsats, because the reconnaissance satellites it was building for the Air Force had certain features in common with communications satellites. Lockheed also made rockets and was interested in whether a market would emerge for private companies to provide comsat launching services. Together with General Telephone & Electronics and RCA, Lockheed mounted studies of the business potential of comsat systems. International Telephone and Telegraph, which had received the contract for the ground stations for the Department of Defense's Courier satellite, initiated an in-house study of geosynchronous communications satellites. The Martin Company floated the suggestion that its Titan missile could be modified to serve as a satellite launcher. General Electric appraised a number of possibilities: manufacturing satellites, offering launch services, and participating in a company that would own and operate a communications satellite system. Even the start-up company Space Electronics, a consulting firm that was a spin-off from Space Technologies Laboratories, went out to Wall Street to hunt for capital to finance a comsat system.[10]

At NASA, the attitude of administrator T. Keith Glennan was shaped by several factors. As a matter of principle he was in favor of a privately owned and operated satellite communications system. He also preferred that the research and development on commercial systems be funded by the private sector to the maximum extent possible. But he was sensitive to the fact that there were a variety of views in Congress and the NASA bureaucracy, and that there was intense rivalry among private firms to enter the new business sector. It would be perceived in political and industry circles to be unfair were NASA to pursue policies that gave the powerful AT&T an advantage. Nor did Glennan himself want to see AT&T sew up the field.[11]

In October 1960, Glennan floated a trial balloon. He proposed that NASA buy rockets and make launch services available to private firms at cost. AT&T immediately approached Glennan to see if it could work out arrangements to put up its Telstar satellites. It was willing to consider a variety of schemes, running from complete assumption of costs to cost-sharing with NASA.[12]

Shortly before Glennan made his proposal, the Department of Defense had reorganized its own communications satellite program, folding the Decree program, along with some others, into Advent, an ambitious system of geosynchronous military communications satellites that were to be furnished with cryptographic and antijamming capabilities. At the same time, the Defense Department had negotiated a new agreement with NASA that allowed the space agency to work on active satellites in lower altitude orbits.[13] Glennan therefore decided that in addition to making a Telstar deal, NASA would sponsor a government-funded program on medium altitude communication satellites, with the contractor to be decided by competition. The hope was to keep AT&T in the field, while preventing it from getting a stranglehold on the technology.[14] A request for proposals for this system, Relay, was issued in January 1961.

By this time, the Eisenhower administration had come to an end, and John F. Kennedy was inaugurated as president. On the political utility of a communications satellite system, the Democratic Kennedy administration agreed with its Republican predecessor. Communications satellite research was one part of the space race where the United States was ahead of the Soviet Union. It was a space application where the United States could project a peaceful image, and one that could render concrete aid to those less developed nations for whose support the United States was competing with the Soviet Union.

By late April 1961, furthermore, Kennedy was searching for a riposte to two communist successes: the Soviet Union had become the first nation to place a man in space and the Cuban army had repulsed a U.S.-backed invasion at the Bay of Pigs. Kennedy therefore seized upon space, calling for one program to land U.S. astronauts on the moon within a decade and another to put a communications satellite system into early operation.[15] To achieve the satellite system, Kennedy asked that an additional fifty million dollars be added to NASA's fiscal year 1962 budget.

The Department of Defense Advent program was by now in serious trouble. Defense leaders, who wanted data on geosynchronous satellites, were more than willing to renegotiate with NASA once again and to permit their brother agency to field geosynchronous satellites. By allowing NASA researchers to use the ground stations DoD had built, at least some of the money it had poured into Advent could be salvaged. In August 1961 NASA finally gave Hughes Aircraft Company a contract. The NASA director choose Syncom as the new project's name.[16]

Both the Relay program, where RCA had won the contract, and Syncom represented a shift. During the Eisenhower regime, Glennan had contemplated that much of the money for communications satellite R&D would come from private industry. With Relay and Syncom, however, it was the government, through NASA, that was financing the R&D, even though from the start the results were expected to undergird a commercial endeavor.

NASA in 1961 very much wanted this expanded R&D role. First, the Kennedy administration and James E. Webb, the man Kennedy appointed as his NASA administrator, believed in a bigger role for government than had the Eisenhower administration and T. Keith Glennan. Second, NASA believed its participation would lead to a better system. As *Business Week* reported in March 1961, "NASA feels that it has no ironclad assurance that industry ever will develop such a [satellite communications] system, and that it must remain in the field until the creation of a system is assured."[17] That must surely have been hyperbole, for NASA must have seen that industrial firms were clawing each other to get into communications satellites. But if it was an exaggeration, there was nevertheless a genuine issue behind it: would industry, left to itself and guided by nothing but the profit motive, design the best possible system? This question was the more crucial in that NASA believed it had knowledge and experience in areas of which industry was ignorant, areas like the effects that launching and the space environment would have on the satellites.[18] Third, if R&D were left to the private sector, wealthy companies with well-endowed research laboratories, notably AT&T, might take such a lead that all chance would be lost for a vigorous field of competing firms. Finally, a constituency for in-house communications satellite R&D was beginning to develop within the agency's ranks. In the Langley center Project Echo team, which by late 1960 numbered nearly 200,[19] and at Goddard Space Flight Center, NASA had groups of engineers that wanted to go on doing research in space communications.[20]

Certainly, in the Syncom project the Hughes Aircraft Company was making use of NASA to test concepts for a satellite the company eventually wished to offer commercially. But NASA was also making use of Hughes. It is not just that Syncom enabled NASA to take the bold steps toward a communications system President Kennedy was mandating. It is also that Syncom allowed the agency to work at the forefront of communications satellite research. Each organization was, in effect, enrolling the other in pursuit of its own purposes.

IN LATE 1961 and 1962, the administration and Congress turned to the questions of how a communications satellite system should be owned and operated, and how, and from where, the satellites would be launched. AT&T argued for a private system. "We believe that the commercial application of satellite communications is a job for private enterprise."[21] Government involvement would add an unwelcome layer of politics and nationalism to what should be governed by purely business considerations. Existing communications companies in the United States and abroad already had the laboratory facilities and the technical and managerial resources to create the system, and the swiftest way to advance would be to allow them to proceed on their own. These established companies would also be best able to integrate a satellite system into the existing communications network.[22]

In tandem with this political fight, AT&T argued over the technological merits of the medium altitude satellites it was proposing to field vis-à-vis the geosynchronous satellites Hughes advocated. AT&T pointed out that the distance from earth to geosynchronous orbit was so great that an annoying delay would be introduced into telephone conversations and speakers would hear an echo of their own words immediately after they had said them. Proponents of the geosynchronous system countered that ways could be found to suppress the echo and that far fewer satellites would be needed for a geosynchronous system and ground stations for it would be substantially simpler.

In Summer 1962, AT&T lost the political battle. Congress, after considering a number of bills, settled on a hybrid solution: a for-profit company, the Communications Satellite Corporation (Comsat Corp.), with stock divided among common carriers, equipment manufacturers, and the general public. The corporation board would have three directors appointed by the president, with the rest chosen in such a way as to prevent any single company from dominating corporation policy.[23] AT&T had negotiated an agreement with NASA in July 1961, whereby it would bear all the expenses for having NASA launch two experimental Telstar satellites. NASA had won the quid pro quo of a royalty-free license on Telstar inventions made from mid-May onwards and the right to sell licenses on these inventions to other firms. The first Telstar, launched in July 1962, was a success. Its launch would come to raise in sharpest terms the issue of privately funded versus taxpayer-funded R&D, for Telstar anticipated by five months NASA's Relay, to be launched in December 1962, and achieved in advance many of the results being hoped for from Relay.[24] Nevertheless, the Communications Satellite Act that Congress passed effectively legislated AT&T out of owning and operating a system of its own design and the company now made plans to join the Comsat Corporation.

Of special interest for our purposes are the functions NASA undertook by virtue of the legislation.[25] NASA was made technical advisor to the Comsat Corporation, and it also was to work with the corporation on research and development. This was subtly but importantly different from both Relay and Syncom. In the former projects, NASA, by means of contracts, could be said to be doing generic research that could be of benefit to any firm that would engage in comsat development. In the Comsat Corporation case, NASA was helping a private, profit-making monopoly in a way that would, presumably, increase its profits.[26]

Second, NASA got the job of procuring launch vehicles and conducting launches for the corporation on a cost-reimbursable basis. This meant that the provision of private launch services, something that Lockheed, Martin, and General Electric had all been eyeing, was out, at least for the present. NASA, conceived as a research and development agency, was instead put in the position of running a quasibusiness launch operation. This insertion of NASA into commercial space operations was to lead in two contradictory directions twenty years later. On the one hand, it would provide a template for NASA's attempt to run the space shuttle as a commercial space transportation operation. On the other, it would provide the personnel for a NASA office, the Office of Commercial Programs, dedicated to helping private companies move into space, including those that were trying to compete against the shuttle as providers of space transportation.

THE FIRST Syncom satellite was launched in February 1963. It was successfully placed in geosynchronous orbit, but failed a few seconds later. NASA and Hughes decided a gas tank had probably ruptured, and searched for a stronger tank material.[27] In July, Syncom II, corrected for the tank problem, was launched. This time the satellite functioned brilliantly. Even a problem previously seen as difficult, that of keeping it in a precise position with respect to the ground, was well solved by Syncom II's engineers. The majority of specialists now affirmed that a geosynchronous system would better meet communications needs than a system of medium altitude satellites like Telstar or Relay. Nevertheless, the Comsat Corporation, which faced the immediate issue of which type of system to order, had not yet decided between them. In December 1963, Comsat issued a request to industry for bids for a "basic system" of either medium or geosynchronous altitude satellites. Four bids were received in February 1964. AT&T with RCA, IT&T with TRW (Space Technology Laboratories), and the Philco Corporation all bid medium altitude systems. Hughes bid a geosynchronous system.

Hughes now made a shrewd business move. It proposed to Comsat a Syncom that would be experimental and operational at once. It would provide some real service, but be essentially a test-bed. NASA, which was working with Hughes on the Syncoms even while it served as technical advisor to the Comsat Corporation, advocated for Hughes' proposal. In March 1964, Comsat agreed and signed a contract for two of these satellites, to be called Early Bird.

Hughes also approached the managers of NASA's Apollo program and proposed that Apollo use a Syncom to link the ground stations that would track the Apollo vehicles. Shortly after, in August 1964, the last of the three Syncom satellites made a successful flight. NASA accepted Hughes' suggestion and in 1965 went to the Comsat Corporation to ask that it buy Syncoms from Hughes to provide NASA this communications link. NASA backed up its promise to be a customer for the Syncoms by agreeing to turn over to Comsat a sum nearly equaling the cost of the satellites themselves if it failed to rent space. Comsat agreed, and ordered four Hughes satellites. By this time, an International Telecommunications Satellite Consortium (Intelsat) was in existence, and Comsat had been absorbed into it. The four Hughes satellites that Comsat ordered became known as Intelsat II. The denomination Intelsat I was given retroactively to the Early Birds.[28]

BY THE end of 1965, Hughes Aircraft Company occupied a commanding position in the manufacture of communications satellites. AT&T had left the field, so its manufacturing arm, Western Electric, had not become a competitor. Hughes had built three Syncoms under contract to NASA. It had sold the Intelsat I and II systems to Comsat Corporation. At the end of 1965 an Intelsat III contract went to TRW for a set of satellites that could fly either at geosynchronous or medium altitudes, but Hughes would go on to win the Intelsat IV series. Meanwhile, Hughes had obtained a follow-up to the Syncom contract from NASA, for five experimental geosynchronous satellites that would later be known as the Applications Technology Satellites.

David J. Whalen is the historian who has laid out most clearly NASA's role in determining the industry leader, and the shape the technology took. In his words, "If the government . . . had not intervened, an AT&T-dominated medium altitude system would almost certainly have been launched in the mid-1960s. . . . Government intervention had two main effects. First, the success of the Syncom series and the NASA commitment to the geosynchronous orbit, . . . made geosynchronous the logical choice for a commercial system. Second, the demonstration of the Hughes geosynchronous system gave Hughes an advantage over all of its potential competitors. This advantage grew as NASA persuaded COMSAT to launch a version of Syncom using commercial frequencies (Early Bird) and offered to be an anchor tenant for the Intelsat II series. The NASA Applications Technology Satellite kept Hughes at the top of its technical form, while Intelsat flew the TRW-built Intelsat III series."[29]

This was industrial policy with a vengeance, aimed not only at a sector, but at firms within that sector. It was partly intentional: Webb, and to a lesser degree Glennan, wanted to prevent AT&T from dominating the industry. But in raising up Hughes, NASA appears to have taken a course more accidental than planned. Fortunately, as Whalen has pointed out, there was another maker of industrial policy on the field. This was the Department of Defense, which, with its SCORE, Courier, and Advent contracts, helped some of the firms that would compete with Hughes down the road.[30]

IT IS clear from the case of Hughes' ascendancy that the question of private versus public funding of R&D was intimately connected with questions of self-interest. AT&T was a wealthy firm and could afford to finance communications satellite R&D. Its strategy was to use its own funding with the aim of getting a system in place before its competitors could enter. Hughes did not have AT&T's resources and it needed outside funding. NASA had its own institutional interests. Doing communications satellite R&D would enable it to stake a claim to a role it wanted: that of undertaking R&D on behalf of the commercial space industry just as the NACA had done R&D for the commercial airplane industry. Satellite communications research was highly visible at the time and provided good public relations for NASA. Communications research also allowed NASA engineers to do interesting work.[31]

NASA and Hughes were pitted against AT&T in this struggle, and NASA and Hughes won. We shall see that twenty years later the alignment would be different. Hughes, by then the dominant manufacturer of commercial communications satellites, would line up with AT&T against NASA to argue that communications satellite research be left to the private sector. In its fight to keep the government from giving a leg-up to *its* competitors, Hughes would use much of the rhetoric and invoke many of the same principles that its opponents used two decades earlier.

North American Aviation and Apollo

Syncom provides a case of how NASA and industry participated in the creation of a commercial industry. In contrast, Apollo allows us to study how government and industry divided the labor of creating government hardware. Here, North American Aviation is the cardinal company to look at, for by the end of 1961 it had emerged, somewhat unexpectedly, with the lion's share of Apollo contracts.

Apollo represented an acceleration of a plan that NASA had already been developing, to send a spacecraft with three astronauts flying around the moon during the 1960s, and follow that with a manned landing on the moon in the 1970s. Through April 1961, the plan had not won administration approval. In May, however, Kennedy seized upon it, and converted it into a crash program. Both the circumlunar flight and the lunar landing were to be accomplished before the end of the 1960s. Estimates for the program were twenty to forty billion dollars. It would be one of the costliest projects the federal government had ever undertaken, and most of the money would be passed through to industry.[32]

North American Aviation was interested. Its business in military aircraft was shrinking as the services cut back on aircraft procurement in favor of missiles. Like Hughes, North American had been hard hit in 1959 by the cancellation of its Air Force contract for the F-108 fighter. North American did not have a large share of the missile market. Its Navaho cruise missile was canceled in 1957 and an ensuing contract for the Hound Dog missile had not been large enough to take up the slack. Nor could the company take comfort from the spurt then taking place in the production of civilian airliners. This was the period when the commercial fleet in the United States and abroad converted from propeller to jet propulsion, and U.S. firms dominated the manufacture of these new airliners. But it was Boeing and Douglas, and to a lesser extent Lockheed and General Dynamics, that served this market.[33]

The business North American Aviation aimed for was government contracts for projects at the leading edge of aerospace technology.[34] The company had just completed the NASA–Air Force X-15, and the plane was getting good coverage in the trade press as it underwent a successful series of test flights.[35] North American had placed its Missile Development Division under Harrison A. Storms, Jr., the engineer who had led

the X-15 project at the end of 1960. Storms, who was interested in securing space projects, renamed it the Space and Information Systems Division.[36] In these circumstances, NASA was a natural agency for North
American to woo.

Apollo hardware would have two major parts. One was the spacecraft, which had to house three astronauts and be able to reach the moon
and return safely to Earth. The other was the booster that would place
the spacecraft on its moonward trajectory. The family from which the
booster would be chosen, the Saturn rockets, had been under development since late 1956.[37] The Army Ballistic Missile Agency in Huntsville
had already begun building in-house a bottom stage (the S-I), and, in
spring 1960, the Douglas Aircraft Company was given a contract to build
an uppermost stage, the S-IV.

At the time Kennedy proposed the moon landing, bidding was in
progress for an intermediate stage, called the S-II. North American Aviation's Space and Information Systems Division bid in competition with
three other firms, Convair of General Dynamics, Aerojet, and Douglas.
This contract was to be supervised by the von Braun team at Huntsville,
which by this time had become a part of NASA, and had been reorganized as the Marshall Space Flight Center. North American Aviation was
judged to have a good chance at winning the Saturn S-II contract. The von
Braun group had worked closely for years with another division at North
American, Rocketdyne. Rocketdyne engines had powered the Army Ballistic Missile Agency's (ABMA's) Redstone and Jupiter, and were to be used
in the S-II stage.

As for the Apollo spacecraft, no hardware development was yet underway. Serious study of possible configurations had begun within various NASA centers in early 1960. In October of that year, NASA awarded
three design study contracts, for a quarter of a million dollars each, to
Martin, General Electric, and Convair Astronautics Division of General
Dynamics. NASA saw such study contracts as a way to enhance industry
capability in space technology. But one of the NASA design teams, at the
Space Task Group at Langley Research Center, continued to work on its
own design.[38] Langley felt heavily the responsibility of safe return for the
astronauts, and followed a conservative approach, one that made the
Apollo capsule an expanded and modified version of the Mercury. Anyone tracking events would have had plenty of reasons for intuiting that
the Langley design would prevail over those of industry. For example, at
a February 1961 review of the studies by Martin, Convair, and General
Electric, Max Faget, a chief designer for the Space Task Group, showed
his irritation that none of the three contractors had thought to pursue the

design route the Space Task Group was traveling and submit an Apollo design based on the Mercury capsule. His comments were taken by the companies as an invitation to reorient their work.[39] And, indeed, when NASA prepared to issue a contract for the spacecraft in July 1961, it sent out a request for proposals based on the Langley design.[40]

At North American Aviation, Space and Information Division president Harrison Storms wanted to bid on the Apollo spacecraft. Here, however, North American Aviation was thought by the industry to be on shaky ground. Its four major competitors were McDonnell, Martin, Convair-General Dynamics, and General Electric. McDonnell was the only company in the United States with hands-on experience in building spacecraft. It was NASA's contractor for the Mercury capsule. In mid-1961, it was also working with NASA to design a follow-on Mercury Mark II, a ship that would evolve by year's end into the Gemini spacecraft.[41] Martin, Convair-General Dynamics, and General Electric were the three companies that had won the NASA contracts for design studies of the spacecraft. Each had used its contract as a buy-in. Although the contracts paid only a quarter million dollars, Convair-General Dynamics had spent one million dollars on the work, General Electric had spent two million, and Martin had spent three million.[42] A firm can develop a lot of expertise for three million dollars!

As each of the five companies began feeling out other firms with the idea of enlisting them as teammates, North American went to Chance Vought, where a group had been working on designs for lunar spacecraft since early 1960. But Chance Vought preferred McDonnell; that company looked to them like the obvious winner. "I never will forget our smugness in sending [North American] packing," recalled one of their engineers seven years later, "we felt that they were sure to lose this and had nothing to bring to the party."[43]

North American had turned in an "arrogant" proposal for the Mercury capsule.[44] For the Apollo bid, Storms determined instead to be completely "responsive." North American would follow the design laid out in the request for proposals, building its case not on innovative design ideas, but on the company's superior ability to execute NASA's intentions. Storms evaluated Faget to be "an egocentric genius with very definite ideas about what the end product should look like. Max had personally laid out the lines of the Mercury capsule and jammed the design down everybody's throat, and there was every indication he would do the same thing on Apollo." Storms therefore wanted to take NASA's "vision and reflect it right back as faithfully as a mirror."[45]

NASA's process for selecting a contractor had two stages. First, a source evaluation board would be established: it would develop criteria for rating the proposals and rank them. Then the top three officers, Administrator Webb, Deputy Administrator Hugh L. Dryden, and Associate Administrator Robert C. Seamans, Jr., would meet with the source evaluation board, review its criteria and ratings, and make a final decision.

While North American worked on the spacecraft proposal in September 1961, it learned it had won the contract for the S-II Saturn stage. The lore was that NASA would not give more than one major contract to any single firm. The agency was seen as wanting to spread its contracts as widely as possible, both in terms of companies and geographic locations. This would help ensure a cadre of capable firms for future contracts, and also build political support for its expanding space efforts. Despite having been awarded one contract, North American's Space and Information Systems Division continued its effort. It employed three tactics. It worked very hard; it put in the best talent and the most resources—about five million dollars—that it could muster; and it asked a modest price.

In November, the Spacecraft Source Evaluation Board reported to NASA top management. It placed Martin's bid, also technically "responsive," on top and recommended North American Aviation as second choice. Webb, Dryden, and Seamans considered the board's recommendations and decided to go with North American. The three men wanted a company versed in building piloted craft to make the capsule for the Apollo, rather than a firm like Martin, whose focus was on missiles. They admired the performance of North American's X-15, and knew that many of the engineers who had worked on that plane had been brought over to work on the Apollo spacecraft. Moreover, the Space Task Group that was to manage the spacecraft and some of the astronauts pressed for North American.[46] McDonnell, which in any event had only tied for third place in the source evaluation board's ratings, was out of the question. McDonnell was about to receive a contract for the Gemini project, and NASA was wary of giving any company more work than it could handle. On November 28th, NASA announced the award of the Apollo spacecraft to North American Aviation.

North American was now NASA's largest contractor in dollar terms. It had the S-II stage, the Apollo spacecraft, and through its Rocketdyne Division, a number of contracts for three of the engines for the Saturns.[47]

A controversy over whether to reach the Moon by earth-orbit-rendezvous or lunar-orbit-rendezvous wound to an end after June 1962. In earth-orbit-rendezvous, a series of Saturn launchers would place components of the space vehicle into earth orbit, where they would be assembled and propelled on to a direct moon landing. In lunar-orbit-rendezvous, a

Saturn would launch the spacecraft, equipped with an extra lunar module, into orbit around the moon, from where the module would descend to the moon's surface. First the NASA Centers, then headquarters, and finally the White House, signed on to the lunar-orbit-rendezvous, using a version of Saturn called the C-5. (Later the rocket was renamed the Saturn-V.)[48] This decision both diminished and enhanced North American Aviation's role. It diminished it insofar as the spacecraft North American was building would now not touch the moon, but only orbit around it. A separate lunar module would be built to descend to the moon. North American's Space & Information Systems Division toyed with the idea of bidding on the lunar module, but Apollo managers at Huntsville and Houston (by then the Space Task Group, originally at Langley, had moved to Houston, Texas) warned them not to; the load of work the division had already shouldered was too heavy.[49]

From another angle, however, the lunar orbit decision made North American more crucial. This is because it fixed the Saturn C-5 as the Apollo rocket, and the three-stage C-5 had North American's S-II as the middle stage. By late 1962, then, North American was not only NASA's most remunerated contractor. It was also the contractor that was most crucial for the Apollo program, and the one that would receive the most scrutiny.[50]

Managing North American

In managing Apollo, Webb and the NASA leadership confronted what they saw as the most complex research and development project the United States had ever undertaken.[51] The lunar landing that they had signed onto not only had to be carried out safely and successfully; it had to be done on time, within the decade of the 1960s, to get to the moon before the Soviets. The task was enormous: nevertheless Webb had still other goals. One was to create an industrial and research infrastructure of so high a quality that the United States would become the preeminent nation in space technology.[52] This meant, among other things, drawing in new companies or giving firms already working with NASA assignments in areas in which they had not previously performed. NASA would benefit directly. The greater the number of firms with experience, the more competition NASA could hope for in its procurements. Furthermore, the more widespread expertise was in industry, the more innovations in both the technology and the management of the technology would be forthcoming.[53]

Related to the desire to increase the national competence in space was

Webb's wish that NASA and the Apollo program serve as matchmakers to bring together in regional alliances university, government, and industry organizations to enhance American technological capability. Webb, part of an activist administration committed to economic growth, saw the Apollo program as an engine for domestic growth as well as a race with the Soviets. Getting to the moon would help win the cold war by demonstrating the nation's technological and organizational capabilities to the world. But Apollo could also spread money to less developed parts of the nation, and serve as a template for other large social programs.[54]

Webb also wanted to educate industry on how to profit from the kind of contracts NASA was offering. Aerospace companies were used to making their money on production runs of hundreds of items. It was common practice to buy in by underbidding on development, with the idea of making up the losses in the subsequent hardware contracts. But for the Apollo procurements, the work *was* largely development. The actual "run" might amount to one or two dozen items, and even these would not be standardized but would differ one from the other. Closely connected was the question of assuring reliability. When the government bought hundreds of standardized items, as for example, military planes or missiles, it was reasonable to achieve reliability by destructive testing of tens of experimental articles. This procedure was obviously not available when there were only about a dozen items in all. Furthermore, each of these items, be they spacecraft or rocket stages, would be so monumentally costly that it would have been insanity to deliberately try to destroy it. "Quality assurance" had to be done through attending to the manufacture and inspection processes with meticulous care.[55]

To the Apollo program and its ancillary goals, Webb brought some well-thought-through ideas on the distribution of roles between government and industry. Most of the hardware and services for Apollo were to be provided by private industry. Webb saw the private sector, properly used, as a way to provide the government with resources in a flexible manner, as well as an engine for achieving efficiency, rapid results, and low costs.[56]

Brains were part of those resources. Webb thought it essential to bring "from the beginning the best brains of these important American companies like Boeing, Douglas, North American, Grumman" into the design of a piece of equipment.[57] On the other hand, Webb believed that national space policy should not be turned over to private firms. It was government acting in the public interest that had to determine what should be done, when it should be done, and for how much money. Furthermore, it was vital that government retain the technical competence to assure that its contractors perform well, that their costs were justified, and their scheduling realistic. It would be intolerable if NASA monitors had to fall back on mere compliance with fiscal or administrative rules to evaluate contractors.[58] NASA needed to maintain what the aerospace community calls, in a curious inversion of English usage, "visibility," that is, detailed knowledge of practices and outcomes at the contractor's facility.

These goals and principles stood in an uneasy tension with some of the realities NASA faced. For example, NASA was expanding rapidly. Total agency personnel went from about 10,000 in mid-1960 to 17,500 in mid-1961, 24,000 in mid-1962, and 30,000 in mid-1963.[59] Even before Apollo took center stage in May 1961, there were differences among the capacities of various centers to oversee the production of hardware. Marshall Space Flight Center engineers had manufacturing experience and were well equipped to monitor production. Engineers at the Manned Spacecraft Division at Langley (soon to become the Manned Spacecraft Center at Houston) did not have that experience. Now, with the pell-mell growth of these centers, people of uneven quality and training were pouring in. It was an open question, therefore, whether NASA had the technical and managerial competence that Webb saw as essential.[60]

The high visibility of Apollo in the United States and abroad conflicted with its strict schedule. The NASA and industry engineers involved were eager to work to the highest standards. "Working on the moon program," recalled one North American manager, "was like building the pyramids . . . people think: . . . The sky's the limit—Nothing is too good for it."[61] Such striving for excellence is a notorious brake on getting a product out the door. Of course, it is the task of management to prevent the engineers' penchants for improvements from running wild. But technical management was still primitive at NASA in 1962. McDonnell's contract for the Mercury capsule had been managed informally and improvements had been instituted without many bureaucratic structures in place to guide them. Apollo was still more important a project and the urge to make it as good as possible was still stronger.

Webb's desire to build a preeminent infrastructure in space research and manufacturing for the nation brought with it the danger of wasting time and money. It meant giving contracts to firms with less experience, firms that needed periods of learning. It meant research grants to a wide number of university groups, not all of them of first rank. There was a way in which all this could aid the Apollo program. Crash program that it was, Apollo was nevertheless going to stretch over nearly a decade. In the United States, it is difficult to maintain consensus either among lawmakers or the public for a costly program with such long duration. Webb therefore had to build a constituency, and bringing in as many universities and firms as possible was one way of doing this. Contracts make constituents out of both the organizations that receive the money and the Congresspersons that represent them. The goal of expanding U.S. capabilities in space technology was therefore as much a matter of political prudence as it was of nationalistic vision.[62] But on the other hand, spreading money around could divert resources from the tasks more immediately related to getting Apollo off the ground. That meant that using Apollo as a means to build U.S. expertise in space was not entirely consonant with carrying out the project itself. Recovering the resources so as to consecrate them to the near-term goal of landing on the moon was, by the same token, made doubly difficult precisely because "building expertise" had political as well as idealistic roots. Such conflicts and contradictions, inherent in the goals for Apollo verses the realities that attended it, would also affect NASA's relations with its North American contractor.

FEDERAL AGENCIES intrude into the activities of their private-sector contractors far more than the uninitiated might intuit, given the free enterprise structure of the U.S. system. NASA's intervention in North American's Apollo work probably went as far as such intrusion goes. This was not only, or even chiefly, because North American had such a large piece of Apollo, with responsibility for the engines, and both the second stage of the Saturn rocket, and the command and service modules. Rather it was that NASA saw North American's Space and Information Systems Division as its most troublesome contractor. Where there was trouble, NASA was inclined to send in its personnel and its techniques, and it had the engineering and manufacturing expertise to make this laying on of hands possible.

Even without special problems, NASA had, from the start, expected to participate rather fully in the work. The overall, rough design of the Apollo capsule had been a NASA product. The Space and Information Systems Division fleshed out the design details in constant interaction with NASA and subject to the approval of its army of monitors.[63] Choosing subcontractors began with NASA and North American sitting down together to negotiate lists of approved firms; only then did North American talk to the companies on the lists to determine where the best deals lay. Testing of systems and subsystems was carried out at both NASA and company facilities, by both NASA and company personnel.[64]

North American did enjoy unusual latitude on the design of the S-II stage. Marshall worked cheek-by-jowl with Boeing in designing and manufacturing the first Saturn stage, the S-IC. The upper stage, the S-IVB, also evolved out of Huntsville designs. But for the S-II, the Space & Informations Systems Division brought some radical ideas of its own to the table. The division had committed itself to building a structure whose "dry weight" (before the addition of propellants) would be a mere 7 percent of its total weight (structure plus propellants). North American therefore proposed having the fuel and oxidizer tanks share a common partition, in the form of a dome or bulkhead. This would reduce the stage's length below what it would be if the two tanks were separate and thus decrease the weight. The bulkhead would form a complex curve to allow it to withstand the forces of the propellants. Shaping its segments, welding them together, and keeping the welded dome—in some places as thin as 0.79 millimeters—from sagging, would call for creative methods of fabrication.

Huntsville engineers were skeptical of this suggestion. The common bulkhead that was being used in the S-IVB stage was only 22 feet in diameter. The S-II bulkhead would have a 33-foot diameter, close to the practical limit for such structures. Nevertheless, Marshall let North American proceed, specifying only that the company work up some parallel conventional designs as a back-up.

Another part of Space and Information Systems Division's weight reduction strategy was to make the tank walls of the aluminum alloy 2014T6, a material that gets stronger as it gets colder, so that it would be at its most robust when in contact with the liquid hydrogen and oxygen that were to fill the tanks. It meant that the insulation for the tank would have to be on the outside and would be in contact with metal that became super-cold, once the propellants were loaded. Maintaining the insulations bonding to such cold surfaces created additional problems for North American's production engineers. So did the fact that alloy 2014T6 was exceptionally difficult to weld.[65] But Huntsville, again, allowed North American to proceed.

North American initially had full control over its staffing. Division President Harrison Storms had chosen John W. Paup to lead the Apollo spacecraft proposal effort because he was a good salesman and a hard driver of the proposal team. After Space and Information Systems Division won the contract in November 1961, Storms kept Paup on as spacecraft manager. He had chosen William F. Parker, who had been the chief engineer of the Missile Division out of which the Space and Information Systems Division arose, to lead the proposal team for the S-II and had kept him also as S-II program manager.[66]

Problems were implicit from the very start, however. In late 1960, when Harrison Storms took over what soon became the Space & Informations Systems Division, it had one major contract—the Hound Dog missile—and about 7,000 workers. As the Space and Information Systems Division first competed for, and then won the Apollo S-II and command and service modules, it began to hire, building up to about 10,500 workers by the first half of 1962, and to about 30,000 by mid-1964.[67] This was a fourfold expansion over four years: the new employees were necessarily of uneven quality and the managerial tasks needed to oversee them were heavy. At the same time, Kindelberger was ailing. In early 1961 he gave over the reins of North American to Atwood, and in 1962 he died. Kindelberger has been portrayed as a blunt and forceful leader and Atwood as both more gentlemanly and less decisive. Atwood was inclined to give the divisions the maximum possible degree of autonomy.[68] Thus in its period of turbulent growth, the Space and Information Systems Division was in the position of receiving less rather than more corporate oversight.

There were problems, also, with NASA's management of Apollo. At headquarters, D. Brainerd Holmes, head of the Office of Manned Space Flight, was feuding with Administrator Webb and going behind his back to obtain extra funding and priority for Apollo.[69] At Houston, because the staff from the old NACA lacked the experience and sometimes the temperament to oversee contractors, the center director brought a Convair engineer, Charles W. Frick, to head the office overseeing the Apollo spacecraft. But Frick fit badly into the Manned Spacecraft Center and did not get on with North American's John Paup.[70] Meanwhile, design changes were being introduced into the spacecraft left and right as engineers at NASA, North American, and North American's subcontractors each sought to make their piece of Apollo better. Control of changes was urged by North American and NASA and attempted early on by NASA. The system the agency put in place was ineffective, however, and since changes in one subsystem necessitate changes in others, these uncontrolled alterations were creating chaos.[71]

In 1963, NASA began to get its Apollo management team in order. Holmes, in Washington, and Frick, at Houston, were forced to resign. The Manned Spacecraft Center now put pressure on Storms to replace John Paup. The spacecraft was behind schedule and overweight, and Houston wanted someone in charge at North American who would be less bellicose and more effective. Storms resisted on the ground that personnel was the company's business: "his attitude was that you're buying the product, not the employees."[72]

In late 1963, NASA leaders brought in TRW executive George E. Mueller to head the Office of Manned Space Flight. Mueller, in turn, introduced into NASA about two dozen new managers, many of them Air

Force people with experience in managing large weapons projects. To head the Apollo program office in Washington, he enlisted Air Force General Samuel C. Phillips. He sent Joseph F. Shea, a systems engineer, from headquarters to Houston to take over the Apollo Spacecraft Program Office Frick had headed. At North American, in the spring of 1964 under pressure from his own managers as well as from Houston, Storms finally replaced John Paup, substituting Dale D. Myers. Myers and Shea respected each other and worked well together. Shea also instituted a more efficient system to control design changes at the end of 1964.

Despite these changes, NASA monitors were dissatisfied with North American's performance on the S-II stage and the command and service modules. Welds on the S-II stage were failing tests, and the insulation was not bonding properly to the propellant tanks. In September 1965, a stage built as a test item ruptured when it was stressed to 144 percent of its expected normal loading. With its loss, the testing program slipped behind schedule.[73]

As for the spacecraft, hardware was leaving the factory late. Engineering drawings were behind schedule. The command module was overweight and in addition NASA engineers worried that several dozens of its components were unreliable. Changes continued to be introduced, some clearly necessary, to cope with problems that were emerging, but others superfluous. Subcontractors were late, or were delivering poorly engineered systems.[74]

At the same time, the national consensus that had so strongly supported Apollo in 1961 was weakening. Opponents appeared who argued that space weapons were more important than a manned space spectacular, or that instead of competing with the Soviets, the United States should join them in a joint moon landing, that Apollo was starving more worthwhile science, or other more socially useful projects.[75] For Apollo managers, it became more essential than ever to meet milestones so as to keep the momentum of the program going.

In 1965, Phillips decided to apply to North American a procedure often used in the Air Force. He put together a "tiger team" to visit North American to scrutinize its management practices and its engineering, manufacturing, subcontracting, and quality control.[76] Tiger teams had been invented as a means to go outside normal procedures in problem situations and to get the attention of a company's upper management.[77]

Phillips presented his team's findings in a memorandum to Mueller in mid-December. He savaged Space and Information Systems Division's management and the Division's work on Apollo. "I consider the present S&ID management so weak and ineffective and so far past the point of no return that I am convinced they do not have the potential to meet our requirements." He accused the corporate office of "playing an essentially passive role" except on financial matters. He indicted North American for inadequate engineering, poor fabrication quality, faulty inspections, and cost escalation.

Going further, "[n]ot withstanding the fact that details of organization and assignment of specific individuals are prerogatives of responsible management," he laid out an extensive series of recommendations for changes in North American's personnel and organization. Phillips wanted Storms removed from his post as president of North American's Space & Information Division. He listed by name nearly a dozen others he thought should lose their jobs, including Paul Wickham, the chief engineer for the S-II stage, and Gary Osbon, chief engineer for the command and service module. He suggested, "Streamline the S&ID organization that has mushroomed to the strength of nearly 37,000 . . . so that they can produce better results, faster, and at less cost." He specified some of the programs he thought should be excised from the division and gave detailed advice on reorganizing other programs.[78] All-in-all, the closely typed pages of Phillips' memorandum constituted an example of the kind of incursion the government was prepared to make into the internal arrangements of a private firm when contract work appeared to be going badly.

North American President Atwood went to the Space and Information Systems Division for a detailed inspection. Himself a structural engineer, Atwood paid particular attention to the engineering of the S-II. He concluded that the Phillips assessment had been oriented too much toward schedule and cost, and took too little account of performance. In particular Phillips had not factored in North American's singular success in keeping the dry weight of the S-II stage down. At this point it was at 7.6 percent of total weight, close to what North American had promised in 1961. In contrast, the dry weight of the S-IVB stage that Douglas Aircraft Company was building was 9.6 percent of the total weight. For Atwood, the S-II stage was a triumph of structural engineering and innovative manufacturing. It was also crucial to the success of the entire Apollo mission. The lunar module had been getting heavier and without the savings North American was achieving in the S-II stage, the weight of the upper stages and spacecraft would have been too much for the Saturn to place in lunar trajectory. Much of the lateness of the S-II stage, Atwood believed, could be laid at the door of the demanding objectives North American had had to achieve.[79]

Given their achievements, Atwood thought, it would be ridiculous to remove either Storms or Wickham. He and Storms did reassign people.

William Parker was replaced by Robert E. Greer as program manager for the S-II. To complement Myers, the spacecraft program manager, who was well-liked at NASA, Atwood sent George W. Jeffs, the corporate director of engineering, down from headquarters. The division also began the process of divesting itself of programs and reducing the number of personnel.

To deal with some of the problems that still plagued the welding of the S-II stage, NASA invoked one of its standard trouble-shooting techniques, and set up a joint Marshall Space Flight Center–Space & Information Systems Division team cochaired by one senior production engineer from Marshall and one from North American. The "joint approach . . . was reflected down through the ranks, with contractor and NASA technicians working shoulder to shoulder in searching for answers." The team decreased the humidity in the workroom in which the S-II stage was being assembled, imposed clean room standards, and substituted for North American's procedure of moving the welding tool around the tank, a Marshall technique of moving the parts to be welded past a stationary tool.[80]

Through 1966, things improved. The state of the S-II stage continued to trouble Marshall Space Flight Center managers, but the command and service module, under Myers and Jeffs, was making some progress. NASA was beginning to feel more cheerful about its chief contractor.[81]

During 1966, North American's leaders also entered into negotiations with Pittsburgh-based Rockwell-Standard, manufacturer of automotive parts, aircraft, and other commodities related to machinery or metal-working. For nearly a decade, Atwood and other company executives had wanted to leaven the company's product line—then composed almost entirely of items for NASA and the Defense Department—with commercial goods. As their business moved from large orders of military airplanes to research and development on projects like Apollo they found it increasingly difficult to make profits that would give them the capital they needed for financial emergencies. North American thus sought to diversify by forming a conglomerate with Rockwell-Standard. It was a path that had been marked out by Ramo-Wooldridge, which joined itself to automotive parts manufacturer Thompson Products in 1958 to form TRW, and the Martin Company, which had merged with American-Marietta, manufacturer of construction materials, in 1961, to form Martin Marietta.[82]

The Apollo Fire

These hopeful progressions of events were marred in January 1967 by a sad accident. During a routine test of the Apollo stack on the launch pad, a fire broke out in the command module and the three astronauts inside it were asphyxiated. NASA and North American now had new, unlooked-for, things to share between them: investigating the accident's cause, apportioning the blame, and restoring confidence in and support for the Apollo program.

The investigative panel consisted of representatives from NASA, the Air Force, academia, and for a while, North American. The actual work of examining the wreckage was done by an army of more than one thousand, many from North American, about six hundred from NASA. The panel concluded that the fire probably had begun with a spark, perhaps in wires that had been unwisely placed in a position in which their insulation had been abraded by the movement of a door. The spark had then ignited flammable materials that had, again unwisely, been introduced in excess quantities into the capsule. And because the tests were being conducted in an atmosphere of pure oxygen at a pressure somewhat above that of sea level, the fire had almost immediately blazed out of control.[83] North American was faulted for placing wiring where it could be abraded. NASA and the astronauts were faulted for introducing too many flammable materials into the capsule, some of them astronauts' personal items. NASA was faulted also for choosing a pure oxygen atmosphere, instead of a less flammable mixture of nitrogen and oxygen.[84] Which of these causes each participant in the Apollo program emphasized tended to depend on that person's organizational affiliation.

Webb was furious. He felt let down by both his own managers and those at North American. He now bent all his energies to salvaging the Apollo program. It was clear that changes had to be made to assuage an aghast public and an angry Congress. Both Joseph F. Shea, the manager of the Apollo spacecraft at Houston, and Harrison Storms were insisting that despite the fire the command module was fundamentally sound in design. NASA removed Shea, and Webb demanded that Atwood fire Storms. Again Atwood demurred. He was mindful of what he considered to be Storms' contributions. In Atwood's view, the cardinal error had been NASA's practice of filling spacecraft with pure oxygen at 16.7 pounds per square inch pressure for checkouts on the launch pad. Under such circumstances, any source of ignition was bound to lead to catastrophic results.[85] Webb, a consummately effective Washington bureaucrat, thought Atwood did not appreciate sufficiently the importance of changing the appearance of the program, as well as its facts. It was a stubborn and emo-tion-laden argument. Webb even threatened to place the Apollo contracts with another company although doing so would jeopardize the goal of landing on the moon before 1970. At the end of April 1967, faced with this danger, Atwood acquiesced and removed Storms.[86] At about this time and in the midst of the unfavorable publicity generated by the fire, North American merged with Rockwell-Standard, to become North American Rockwell. In a symbolic sense, it could not have been a more appropriate moment for the company to make itself less dependent on government contracts.

THE REBIRTH of NASA confidence in North American was a matter of both real and symbolic changes. Storms' leaving, in itself, removed an ir-ritant, because top managers at NASA had viewed him as ineffective and abrasive. To replace him, Atwood reached for one of his most senior peo-ple, William B. Bergen. Bergen, a former president of Martin, had gone to North American to be a corporate vice president in charge of Rocket-dyne and the Space & Information Systems Division. After the fire, he agreed to a demotion to Space & Information Systems Division president to help restore the program. At the same time, North American removed information activities from Space & Information Systems Division, and renamed it the Space Division. It was a way to emphasize the priority that the division would henceforth give to NASA systems.[87]

The management changes already made in 1966, together with the new managers Bergen brought to the division and the reorganizations he instituted, made a decisive difference for North American Rockwell's work on Apollo. So too did the fact of the fire itself. It showed that nitpicking attention to the hardware was warranted, that space equipment needed a level of reliability and attention to detail beyond even the standards that had been established for aircraft.[88]

Bergen appointed a separate manager for each spacecraft. He also sent a new leader to head the North American Rockwell (NAR) team at the Kennedy Space Center. These managers began to exercise tighter su-pervision down the line. Bergen himself took to visiting the shop floor at odd hours of the day and night as emblem of the fact that top manage-ment intended to have command of every detail of the work. Programs aimed merely at improving engineering or advancing the state of the art were eliminated. These projects had served Webb's goal of expanding the nation's space expertise and they had been supported, as means to that end, by Harrison Storms. But they conflicted with the immediate task of ensuring an early Apollo landing. The number of staff working on Apollo at the Division fell by one-third.

At North American Rockwell and at NASA, managers instituted still tighter control of the technical directives and demands for change that flowed from NASA engineers for the spacecraft. All changes now had to be funneled first through the program officer at Houston and then through the manager of that particular spacecraft at North American Rockwell. North American engineers were made to adhere rigorously to agreed-on procedures, without any creative flourishes.

The command module was redesigned. After that, a team headed jointly by a senior engineer from North American Rockwell and one from Houston was set up to implement the changes and oversee the process right through manufacturing and testing. The intimate collaboration that this team achieved helped change the psychology of NASA-North American Rockwell relations, away from NASA managers simply monitoring North American work with the adversarial relationships this sometimes engendered. Working side by side with their North American Rockwell counterparts allowed NASA managers to learn of problems at the time they were occurring and encouraged a new candor between company and government personnel.

Boeing Aircraft Company also lent assistance. In the aftermath of the fire, Webb arranged a contract under which Boeing would provide NASA with advice on integrating the spacecraft and the rocket. This Technical Integration and Evaluation (TIE) contract was part of Webb's attention to political appearances. He needed to demonstrate that NASA was doing everything possible to put the program right. But if the Boeing contract served a political end, Boeing also was able to provide concrete help to North American.[89]

By the middle of July 1967, NASA officials had begun to feel more comfortable with North American. In May 1968 the first of the command and service modules made under the new regimen was delivered to the Cape and proved to have few errors.[90] And by late 1968, North American Rockwell and NASA shared the triumph of the first, successful, Apollo flights.

Conclusion

The Hughes and North American cases demonstrate how far the creation of space technology in the 1960s deviates from the simple case in which technological innovation is carried out within a single corporation. A model that incorporates government, but merely as funder and market, is also inadequate. In the U.S. space program, the government participated on every level of the innovation process.[91] It provided engineering designs. It intruded into personnel decisions and organizational arrangements at the firm. It helped choose and monitor its contractors' subcontractors. Its engineers formed part of integrated company-agency teams that worked the problems.

One is tempted to see a breaking down of the boundary between public and private, particularly in the case of Apollo. Here was an overriding national objective and it seemed appropriate to contemporaries that organizations sacrifice some of their individuality in order to meet it.[92] But the conclusion that private and public somehow merged would be an error. The public-private boundary continued to exist by virtue of the divergent institutional goals of the companies and NASA. In manipulating personnel at North American's Space & Information Division, for example, NASA's purpose was to get hardware out on time and up to specifications. In contrast, the Space & Information Systems Division's goal was to build a strong organization, one competent not only to fabricate the S-II stage and the Apollo spacecraft, but to win and execute future contracts.[93] In developing Syncom, NASA sought to enhance U.S. industrial capacity across the board, to carry out a presidential mandate, to participate in a fascinating new technology, and to carve out a mission as R&D arm to industry and the military. Hughes' goal was to develop a line of profitable satellites and enter the market with them.

In the Apollo program, one of NASA's overt aims was the expansion of expertise within U.S. industry. Webb saw this as two-pronged: building technological expertise and developing managerial know-how in running complex projects. Yet, because space technology was in its infancy, NASA itself would be learning as it went along. Moreover, NASA's management procedures were also in a state of formation. What, then, did it mean to say that the Apollo project transferred space expertise to industry? Who learned from whom, and what was learned?

Here the two cases we have considered stand in contrast. Hughes came to NASA with a well-developed concept for a light geosynchronous communications satellite. One guesses that NASA learned quite a bit from Hughes about how to build a satellite. Hughes, for its part, must have come away wiser about choosing appropriate launch vehicles and launch conditions. North American, a neophyte in spacecraft, did not bring a design of its own to its collaboration with NASA on the command and service modules, but it did bring original ideas on design, materials, and manufacture to the S-II contract.

North American got the benefit of plenty of teachers: dozens from NASA and additional personnel from contractors like General Electric and Boeing. Aerospace executives differ in their views on the value of such close supervision. Some point out that deep involvement by a technically competent customer can lead to better products. Others maintain that it slows down work and creates additional red tape.[94]

On the Apollo project, the disadvantages of micromanagement were compounded by the facts that many of the managers were new recruits, lacking technical background, and that the pressure of schedule was enormous. The consequence was that rules were sometimes applied by the managers in a mechanical fashion. The destruction of the S-II test stage in September 1965 may well be an example. Here the rule called for testing the stage at forces up to 150 percent of the stresses to be expected in a real lift-off. North American managers asked that the test be limited to 140 percent of stress. The 150 percent figure was a convention, left over from earlier years when airplanes had to be designed without the benefit of computers and advanced knowledge of the mechanics of fracture. NASA managers denied the request and the stage ruptured at 144 percent, with grave consequences for the Saturn schedule.[95]

On other occasions, the company profited from what NASA had to offer. The lessons that Marshall Space Flight Center taught North American about welding contributed to the success of the S-II. The company learned from NASA a new level of attention to detail and to quality. More generally, it appears that space work brought the entire industry to new heights in quality control. Unforgiving as airplanes are, space vehicles are worse. It is seldom possible to recover the remains of a failed space vehicle and investigate the causes of the accident, as one can investigate a plane crash. For space hardware, failure mechanisms have to be understood, and corrected, beforehand.

On the managerial side, Apollo threw North American into a project more complicated than any that the company had carried out before.[96] On the spacecraft alone, the company had hundreds of subcontractors and thousands of vendors to oversee. North American Aviation's parts of Apollo had to fit into parts being produced independently by other prime contractors. The level of detail and coordination that had to be managed was unprecedented. The reformations that NASA pushed on the Space & Information Division management through the Phillips tiger team and in the wake of the Apollo fire seem to have been of real benefit.

But was this a case of North American learning from NASA? At the time NASA was learning its managerial techniques through the tutoring of borrowed Air Force officers and from companies like General Electric and Boeing. Perhaps the best way of putting it is to say that Apollo did occasion a real extension of space expertise in industry, but it was not simply a flow from NASA outward. Rather, NASA's Apollo opened a network of conduits by which technological and managerial knowledge moved with new facility among NASA, industrial companies, and the military. At the end of the process, North American won a place in NASA's inner circle of contractors.

In contrast, Hughes moved out of that circle. Emblematic is the long legal battle between Hughes and NASA, begun in the mid-1960s, over the patent for orienting the Syncom satellite. Hughes first filed for the patent in 1960, claiming it on the basis of a laboratory model built before the company received its NASA contract in August 1961. NASA claimed that the idea, which used a device that was activated in the demonstration model by electric lights but in space by the sun, could not have been reduced to practice until the Syncoms flew in 1963, so the patent belonged to the government. Hughes took the position that it had contracted with NASA to deliver a satellite, not to surrender its technology. NASA believed it had a duty to prevent Hughes from getting a lock on communications satellite technology, just as earlier it had seen itself obligated to forestall an AT&T monopoly.[97] It seems unlikely that Hughes would have persevered in this legal struggle if the company had been as dependent upon NASA's business, and hence its goodwill, as was North American. Instead of NASA, Hughes was turning more to the military and commercial satellite markets.

In the end, both companies fared well. Hughes won its patent fight and with it substantial royalties. It also retained its dominant position in sales of communications satellites. North American Rockwell surmounted a trying decade so well that, at the end, it was in position to win the biggest NASA plum of the 1970s, the shuttle orbiter.

The Space Shuttle

THE YEAR 1969 WAS A WATERSHED FOR NASA. Outwardly, the agency was brilliantly successful. The *Apollo-8* spaceship had circled the moon at the end of 1968, and its astronauts, Frank Borman, James A. Lovell, and William A. Anders, had become the first human beings to see the moon's far side. Two more successful preliminary Apollo flights followed in March and May of 1969 and in July, the Apollo lunar module landed, and Neil A. Armstrong and Edwin E. Aldrin walked on the moon's surface. NASA had fulfilled the pledge of a martyred president, demonstrating to the world the superiority of U.S. technology and, by implication, the nation's capitalist economic system.

Yet in actuality, NASA had been sliding down the federal totem pole since 1965. In that year, and despite his personal commitment to the space program, the high cost of the Vietnam war led President Lyndon B. Johnson to accede to cutbacks in NASA's budget. NASA appropriations went from $5.250 billion in fiscal year 1965, in current-year dollars, to $5.175 billion in FY 1966, $4.968 billion in FY 1967, $4.589 billion in FY 1968, and $3.991 billion in FY 1969.[1]

The outlook worsened after Johnson left office and a new Republican president, Richard M. Nixon, took over at the start of 1969. Nixon identified the great manned space programs of the 1960s with his Democratic rivals: in the first instance with John F. Kennedy, who had defeated him in the 1960 presidential election, and after that with Kennedy's successor, Johnson. The budgetary squeeze brought about by the war could make it all the easier for Nixon to cut back on NASA programs. So too could the fact that in the country at large, enthusiasm for large technological projects was diminishing.[2] Despite its spectacular performance, NASA in 1969 was just another federal bureau, reduced to scrapping with other government agencies for its share of a restricted budget.

The year 1969 was also a watershed in NASA-industry relations. It would be astonishing if it had been otherwise. Europe and Japan, revived from the effects of World War II, were challenging the United States in world markets. As a result "foreign competition" was taking its place alongside "Soviet communism" as a bugbear of American society. This inevitably raised questions for the NASA-industry relation. What was the purpose of a space program and a space industry in this new world? Were they to compete with the Soviets for prestige, or with the Europeans and Japanese for markets? Was NASA to continue the great adventure of human exploration and settlement of space, with the lunar landing as the first step and with the aerospace industry as its contractor? Or was it to help this industry, and other sectors of U.S. industry, move into an era of commercial development of space?[3]

Despite the changed and less favorable circumstances, NASA would succeed in getting a new manned project approved by the Nixon administration and Congress. This would be the space shuttle, a largely reusable system, with an orbiter that could spend a week or more circling the earth at a distance of a few hundred miles, carrying out experiments or releasing, repairing, or retrieving satellites. The struggle to get approval for the shuttle, however, was a protracted one, lasting three years. And the craft that was finally authorized was less ambitious than the one NASA originally wanted. It was also compromised by features dictated by the Defense Department, an ally NASA recruited to aid in winning approval. In addition, although the shuttle was seen inside the agency as a step on the road toward human exploration of space, given the new political-economic conditions, NASA felt compelled to justify it in terms of economic competitiveness. The agency therefore represented the shuttle as a cheap way to loft satellites and as an instrument for opening up the industrial exploitation of the low earth orbit in which the orbiter would fly.

Two consequences of these events are pertinent for us. First, the long fight for approval involved a continuing renegotiation of the shuttle's basic features. As a result, the design process for the shuttle differed from that for prior astronaut-carrying vehicles. For the Mercury and Apollo capsules, the basic design had been done fairly rapidly by in-house teams of NASA engineers. This was impossible in the case of the shuttle. Instead, for a variety of reasons, industry engineers contributed many of the ideas the shuttle incorporated.

Second, economic analyses showed that the shuttle would only fulfill the promise of providing cheap access to space if it could fly often with its cargo bay full. This led NASA to try to capture for the shuttle all commercial traffic. The agency started to discourage the fielding of alternate, private sector, space transportation vehicles. It set shuttle prices and flight schedules that would be attractive to operators of commercial communications satellites. It promoted research on new types of industrial systems, like electric generating satellites or equipment for manufacturing in space, that could be put into orbit by the shuttle. In all of these moves, NASA was taking steps that directly affected sectors of the commercial space industry. In other words, NASA's shuttle policies were de facto industrial policies that aided some space enterprises and thwarted others. For all these reasons, the shuttle program, so central to NASA, is the natural lens through which to view NASA's interactions with the aerospace industry in the decade of the 1970s.[4]

Designing the Shuttle

At the beginning of 1969, NASA hoped to parley its Apollo successes into an ambitious program of human flight. Under Thomas O. Paine, who became acting NASA administrator in October 1968, and administrator in March 1969, the agency envisioned a space station circling the earth— initially staffed by a dozen astronauts and building up to fifty and more— and another station orbiting the moon. A mixed fleet of shuttles would support the stations. Chemically powered shuttles would travel from the earth to the station in earth orbit while nuclear-powered vehicles would move between the earth and moon stations. Space tugs would ferry payloads from the earth station into geosynchronous orbit and back. A lunar base would be established and this and the space stations would serve as staging areas for human exploration of Mars in the 1980s. All this would be funded by a NASA budget of six to ten billion dollars a year.[5] NASA projected that the earth-to-low-earth-orbit shuttles would each comprise an orbiter and some kind of booster. The booster would return to earth after it had lifted the orbiter, and the orbiter would return later, after completing its mission. Both parts would then be refurbished and reused. In September 1969, a presidential task group headed by Vice President Spiro Agnew endorsed NASA's complex of space stations and shuttles and recommended the manned mission to Mars.

Even before the task group's report was issued, the Mars mission was severely attacked in Congress.[6] In March 1970, President Nixon replied to the recommendations by way of a statement of his administration's space policy. Space activities, he declared, would no longer be a matter of crash programs infused with massive concentrations of resources. Rather, "space expenditures must take their proper place within a rigorous system of national priorities." Accordingly, he was sending Congress a budget for fiscal year 1971 with a NASA appropriation of $3.3 billion, lower than the $3.7 billion of fiscal year 1970.[7] The statement made clear that NASA's vision of orbiting stations, lunar and planetary bases, tugs, and ferries was not going to be implemented. The administration would not even approve an earth orbiting space station. The outlook was more favorable for the fully reusable shuttle, but approval of it was not a certainty.[8]

BY THE time President Nixon took office, studies of vehicles that could shuttle back and forth to low earth orbit had been going on in the United States for more than a decade.[9] Some of the earliest had centered on the Air Force's Dyna-Soar, a glider whose swept-back wings gave its silhouette, seen from above, the shape of the Greek capital letter delta. Dyna-Soar was designed to be launched into orbit vertically by a Titan booster; on reentry into the atmosphere, its shape would generate aerodynamic forces that would permit its pilot to maneuver it to a secure land base of his or her choosing. Dyna-Soar would carry military personnel who could transfer from it to an orbiting space station. Boeing had the major contract for Dyna-Soar, but many other companies held smaller study contracts, including Martin, McDonnell, Lockheed, Hughes, and AVCO.[10] Department of Defense Secretary Robert S. McNamara canceled the Dyna-Soar project at the end of 1963, but other studies on reusable orbiters and reusable boosters continued. Emanating from NASA's Marshall Space Flight Center and the Air Forces's Wright Patterson Air Force Base, they spawned contracts to most of the major aerospace companies, among them General Dynamics' Convair, Boeing, Lockheed, McDonnell, North American Aviation, and Martin Marietta.[11] And as industry engineers accumulated design experience with earth-to-orbit shuttles, they formulated designs of their own.

An example of an industry-generated design is Lockheed's STAR Clipper.[12] Its history begins with Maxwell W. Hunter II, who in 1965 joined Lockheed Missile & Space Company as special assistant to E. P. Wheaton, the vice president for research and development. Hunter was an advocate for inexpensive ways of getting into orbit. In his view, designing a single-stage vehicle was fundamental to getting up cheaply. Multi-

Dyna-Soar and some of the pilots who would have flown it. This manned space glider, never built, featured swept-back wings, angled fins, and a black exterior that would have helped to reradiate heat absorbed in reentry.
Source: NASA Headquarters History Office.

stage systems like the Apollo jettisoned each rocket stage as it completed its propulsion function. Such systems would be costly even if all their stages were designed to be recoverable and reusable. This is because "the stages are strewn about the planet and on orbit . . . [and i]t is difficult to believe that recovery costs can be made anywhere near as low as those achievable by a single-stage device."[13] But a single-stage vehicle would be heavy, and to get it into orbit required energy well beyond that of any chemical engine he could easily foresee. He therefore proposed to make the space vehicle lighter by placing its fuel outside, in an exterior tank. The tank would be an inexpensive throw-away. Instead of a single-stage vehicle that looked too difficult to attain or a multistage, fully recoverable system he thought would be too costly, Hunter opted for what was known as a $1\frac{1}{2}$-stage system.[14] At Lockheed, it was also known as the STAR Clipper.

Hunter and Wheaton conceived of all kinds of uses for such a cheap spaceship. It could serve as a test vehicle for NASA advanced science and technology programs. It could be a manned orbiter for Air Force reconnaissance or an orbital bomber. It could be a platform for the manufacture of materials in space. It could even serve as an excursion craft for jetsetters who had seen and done everything else and wanted to spend twenty-four hours in orbit.[15]

In 1966, John F. (Jack) Milton came to Lockheed from Boeing Aircraft Company. Milton had worked on the Air Force Dyna-Soar project. He took over as STAR Clipper program manager, and under his guidance, the Lockheed design took on some of the aspects of Dyna-Soar, with no wings but with a delta-shaped body. Unlike the Dyna-Soar, which had no power of its own, the STAR Clipper had three rocket engines in the rear. Hunter's extra "½ stage" appeared in the form of a V-shaped tankage, within which the orbiter nestled.[16]

While Milton and his engineers worked out the details of the STAR Clipper, they and Hunter talked it up to the Department of Defense and NASA. Lockheed viewed the NASA center design teams as bent on implementing only in-house design and technology. To get Lockheed concepts into NASA hardware, company engineers thought it necessary to insinuate them at an early stage in NASA's thought processes. Maxwell Hunter recalled that they spent two years trying to get Faget to "invent" the Lockheed design.[17] To Faget's objection that the STAR Clipper relied too much on unknown technologies,[18] Lockheed had a reply. It was that these technologies were unknown to NASA but not to Lockheed, which had studied them and had data on them. In effect, the company was laying at NASA's door exactly the fault Walter Dornberger had warned against in 1958, that of neglecting space technology along the broad front in order to pursue the specific technologies required by its particular projects.[19]

It is clear that industry's position vis-à-vis the shuttle was very different than its position with regard to the Mercury and Apollo capsules. Before the Mercury was designed, it is true, AVCO had already been studying ballistic reentry capsules. It is also true that three companies, Martin, Convair, and General Electric, had been given study contracts to work on Apollo capsule designs. But there was nowhere near the depth and breadth of experience that industry had with respect to shuttle vehicles at the end of the 1960s. And in the earlier projects there had not been industry designs, like the STAR Clipper, with all the tenacious advocacy that design engineers can bring to their own ideas.

THE SHUTTLE also differed from Mercury and Apollo in a second crucial way: its function was not as clear cut. The Mercury project had the

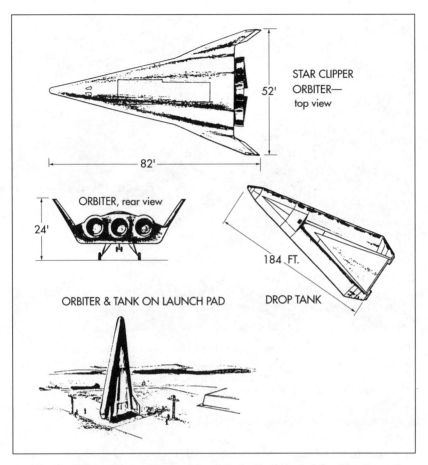

Lockheed's STAR Clipper. The Lockheed design had angled fins like Dyna-Soar and external tanks that enclosed the orbiter at lift-off.
Source: Lockheed Missiles and Space Company, "Briefing to NASA Headquarters, January 6, 1968, Space Transport and Recovery System," courtesy of Lockheed Martin Corporation.

mission of putting an astronaut into orbit as quickly as possible. The Apollo project had an equally well-defined goal of a moon landing before the end of the 1960s. The shuttle was a different matter.

First of all, although there had been some talk of separate NASA and Air Force shuttles during the 1960s, by 1969, NASA and the Defense Department had agreed to a single, national vehicle to serve both agencies. But the Air Force had requirements that were different from NASA's. Chief among them was the military's need for a vehicle so maneuverable

Artist's conception of the Lockheed orbiter and external tank immediately after separation. The sketch shows the STAR Clipper configuration as it might look shortly after the successful jettisoning of the tanks.
Source: NASA Headquarters History Office. Originally published in John F. Guilmartin Jr. and John Walker Mauer, *A Shuttle Chronology, 1964–1973,* Houston: NASA Johnson Space Center, Report JSC-23309, 1988, II-18.

that it could land after only one orbit at a base of its choosing, so as to evade an attacking enemy. The Air Force translated this into a requirement that the vehicle be able to deviate 1,400 nautical miles to either side of its orbital path on reentry. (A nautical mile is about 1.15 times as large as a statute mile.) Such a high degree of maneuverability would impose much higher heat loads, and a shuttle designed for it would therefore need a much greater weight of thermal protection material. Second, the Defense Department needed a large cargo-carrying capacity, enough to carry into orbit some of the big new reconnaissance satellites. It wanted a cargo bay with dimensions of fifteen by sixty feet and a payload carrying capacity of about 50,000 pounds. This was less contentious, because NASA also wanted a bay large enough to bring up parts of a station for assembly in space. A large cargo bay would, of course, also increase the weight of the shuttle.

In addition to differences between the Air Force and NASA there were disagreements within NASA itself. Some in the agency saw the shuttle's main usefulness as an ancillary to a space station, an aid in its construction and a logistics vehicle to deliver and retrieve space station personnel and supplies. Others, however, saw it chiefly as an inexpensive way to reach space. For them, it would serve a multiplicity of functions; releasing and servicing satellites, supporting the station, even transporting passengers via space from one point on earth to another.[20] Chief among the second view's partisans was George Mueller, the powerful head of the Office of Manned Space Flight. Above and beyond these discords were the institutional interests of the centers. Marshall and Houston, in particular, wanted a shuttle that could provide interesting, continuing, work for their engineers.

Given such disagreements over function, it is not surprising that different groups within NASA took different positions on the shuttle's configuration. The Manned Spacecraft Center favored a prototype that could be put into operation quickly so that there would be no gap in the presence of U.S. astronauts in space between the last of the Apollo flights and the new manned system. And Houston was not overly concerned with Department of Defense requirements. Marshall, in contrast, wanted a large shuttle, which would give it scope to develop large, new rocket engines. Headquarters, for its part, was wary of playing around with prototypes, because it did not believe it would be able to secure funding for both a prototype and a final system. Further, headquarters thought that in the political climate that existed, it needed Department of Defense support. NASA, in its view, could not afford to be cavalier about Defense requirements.[21]

MAXIME FAGET, by that date director of the Manned Spacecraft Center's Engineering and Development office, himself offered a design for a small shuttle. Faget did not like the STAR Clipper, for all of Lockheed's proselytizing: he thought its chances for satisfactory performance were marginal. Another industry idea, the Convair-General Dynamics Triamese, struck him as better. But one element of Faget's inventive style was to search out entirely new ways of doing things, so as to complement what was already on the table. By April 1969, Faget presented a design for an orbiter shaped like a fat-bellied, stub-winged airplane. It would ride into space on a much larger, but similarly shaped booster. Both parts would be capable of returning to a landing, and both would be reusable.[22] His initial proposal called for a craft with a 12,500-pound payload capability, and the ability to land about two hundred nautical miles to either side of its orbital path.

The Convair-General Dynamics Triamese. Convair engineers conceived it as an orbiter sandwiched between two reusable boosters, all with identical airframes and propulsion systems. At right rear, the system is ready for vertical lift-off. In the foreground one of the stages, its retractable wings deployed, is landing.
Source: NASA Headquarters History Office, originally published in Guilmartin and Mauer, *A Shuttle Chronology*, p. II-139.

Faget's design patently did not meet the Department of Defense requirement for high cross-range. In June 1969, Wright Patterson Field's Flight Dynamics Laboratory and the Air Force's Aerospace Corporation made a preliminary attack on it. They contended that data for evaluating the concept was so sparse that the Manned Spacecraft Center's confidence in it constituted "an anomaly." NASA's Office of Manned Space Flight had by then accepted the Department of Defense's requirement for a 50,000-pound payload capacity, and the Aerospace Corporation and the Air Force Flight Dynamic Laboratory asserted that one could scarcely evaluate Faget's design usefully until it were scaled for that load. They did not expect that to improve things, however: "Current thought is that . . . the aerodynamic and thermal loads would not improve . . . and that . . . weights would become even more critical."[23]

Faget, in turn, enlisted NASA's Langley and Ames Research Centers to run models of what he was now calling the DC-3 Shuttle in their wind tunnels.[24] At the same time, the Houston center issued a request for proposals for industry studies to increase the depth of analysis of Faget's concept and determine its feasibility. McDonnell Douglas was awarded a contract for the work in July 1969. Shortly after, North American Rockwell, which had a shuttle study contract under Manned Spacecraft Center management, was instructed to devote part of its effort to the DC-3.[25]

In November, the Air Force launched a still more blistering assault. It had carried out wind tunnel tests of the Manned Spacecraft Center shut-

Max Faget's shuttle design. Faget believed an orbiter and booster that resembled subsonic aircraft as much as feasible would simplify the development process and improve the in-atmosphere flight behavior of the shuttle's components.
Source: NASA Headquarters History Office, Photograph 7897.

tle's reentry behavior, and claimed that they showed heating so severe that the weight of material needed to protect the vehicle would be excessive. It said the craft would also be subject to dangerous vibrations. It attacked Manned Spacecraft Center estimates of the cargo weight it could carry as unrealistic and it added an attack on Houston's whole approach to the analysis of the vehicle's structure.[26] Faget's riposte to this was the final report of the McDonnell Douglas study, which concluded that technical analyses showed "this concept is a viable configuration."[27] Meanwhile, Faget could counter the charge of lack of data (made in June by Wright-Patterson and the Aerospace Corporation) with the results of in-house work and the North American Rockwell contract.

While Faget and the Air Force skirmished, aerospace companies were putting forth their own ideas. North American Rockwell advocated an idea widespread in NASA at the start of 1969: making use of Apollo hardware for an interim shuttle system. McDonnell Douglas promoted a vehicle derived from the Gemini capsule it had manufactured for NASA and also an in-house design, the 176H Orbital Support Spacecraft, with a delta-shaped body, expendable fuel tanks, and retractable wings. Lockheed, as we saw, proselytized for its STAR Clipper.[28]

During 1969, new technical information was being developed, much of it from a group of NASA contracts known at the time as the Integrated Launch and Reentry Vehicle contracts, and retroactively as the Phase A Shuttle contracts. The contracts paid $300,000 each to Convair of General Dynamics, North American Rockwell, Grumman, and Lockheed, and were structured to last about six months. Each had a slightly different focus. General Dynamics-Convair, under contract to Marshall Space Flight Center, was to place emphasis on the Triamese, which consisted of two reusable booster stages and an orbiter. The boosters and orbiter were identical in size and shape and the idea was to save money through this commonality of structure. Lockheed, also under contract to Marshall, was going to stress its own STAR Clipper. McDonnell Douglas, to be monitored by Langley, gave special attention to a Langley concept called the HL-10, which had no wings but did have a body shape that gave it lift within the atmosphere. McDonnell Douglas also devoted some of its efforts to its in-house 176H design. North American Rockwell, monitored by Houston, looked especially at configurations that united reusable orbiters with expendable boosters.

In addition to the Integrated Launch and Reentry Vehicle contracts, McDonnell Douglas had a contract with the Manned Spacecraft Center to study a shuttle based on the Gemini capsule and another, as we saw, for Faget's DC-3 shuttle. In-house studies also continued.

Major decisions were being made at headquarters in tandem with the contract work. NASA acquiesced to the Air Force wish for a high cross range. It decided on a fifteen-foot by sixty-foot cylindrical cargo bay. It negotiated a compromise figure with the Air Force for the amount of lift the maneuverable orbiter would have in the atmosphere. Finally, it came down for a fully reusable system. This would require greater sums for development but would allow cheaper operations cost and thus conform to George Mueller's vision of the shuttle as an inexpensive space transportation system.[29]

By the end of 1969, the engineering results and the managerial decisions served to limit the number of contending designs. The Convair Triamese and the Langley HL-10 were ruled out for technical reasons. The ballistic vehicles, derived from Apollo and Gemini, could not be scaled up for a 50,000-pound payload and still be light enough for a parachute landing on land: a recovery from land rather than water was indispensable for low-cost reusability. The decision for a fully reusable shuttle ruled out the STAR Clipper and the McDonnell 176H, because both featured expendable tanks.

It was now time for NASA to invite proposals for more tightly focused design studies of the shuttle. These were the so-called Phase B shuttle studies, for which NASA released a request for proposals in February 1970. The request called for a fully reusable system comprising a booster and orbiter. The system was to be launched vertically, and both stages had to be capable of landing horizontally. Interestingly, although headquarters had already committed NASA to the Air Force requirement of high cross range, and although the DC-3 shuttle design had strong opposition at headquarters, Langley Research Center, and elsewhere within NASA, the request for proposals called for study of configurations for both two hundred and fifteen hundred nautical miles. It was an illustration of how hard it was to close a debate within a complex organization like NASA, as well as the clout Max Faget had within the agency.

NASA received four proposals. Boeing and Lockheed headed one team, with Boeing (listed as prime) designing the booster and Lockheed the orbiter. North American Rockwell (prime, vehicle) with Convair-General Dynamics (booster) was the second team, and the third was McDonnell Douglas (prime, vehicle) with Martin Marietta (booster). Grumman headed its own team.

In June 1970, NASA selected the North American Rockwell team to work under contract to Houston, and the McDonnell Douglas team to work under contract to Marshall. Each contract was worth eight million dollars. Grumman, which did not want to be disadvantaged in the coming competition for the production contracts, persuaded NASA to award it (with Boeing as teammate) a consolation contract of four million dollars to look at alternatives to fully reusable systems. Jack Milton, who was still battling for Lockheed's semireusable STAR Clipper, won a one million dollar contract to continue design studies. Chrysler got three quarters of a million for a heterodox, fully reusable scheme.[30]

In-house studies were proceeding as well. Marshall Space Flight Center was studying a shuttle that used one of their Saturn rockets as booster. Faget had just negotiated a contract with the firm of Ling-Temco-Vought (LTV) for a series of DC-3 shuttle studies, which had the effect of extending his in-house staff by adding LTV's Vought Division engineers.[31]

GEORGE MUELLER left NASA at the end of 1969, and in January 1970, Dale D. Myers, of North American Rockwell, took over NASA's Office of Manned Space Flight. From this position at headquarters, Myers was acutely aware of NASA's need for Air Force support and of the consequent importance of satisfying the Air Force requirements. By mid 1970, he had a chance to examine the DC-3 shuttle design and he did not like it. It had a tail, which Myers thought ill-advised for a plane that had to fly part of the time at hypersonic speeds. The amount of heating different parts of the craft would suffer on reentry was overly sensitive to the angle at which it would enter the atmosphere. It had straight wings. Myers, who had worked with delta-wing aircraft for the Air Force believed that a delta-wing configuration was needed to satisfy the Air Force requirement for cross-range, and thereby to secure Air Force support in selling the shuttle to the administration.[32]

Myers pushed the centers to devote less of their resources to in-house studies and to place more reliance on the design studies of contractors. In June 1970, he wrote to Robert Gilruth, director of the Houston center and Eberhard Rees, director at Huntsville: "The contractor effort should be the focus for the ultimate product. I hold you responsible to limit the in-house studies . . . to activities which truly supplement and support the industrial effort."[33]

In an interoffice memo, Rees commented tartly "Dale Myers is somewhat allergic to 'too much' government interference."[34] This was true enough and Myers had several reasons for his "allergy." First, the pressure on the budget bespoke the need to bring down NASA's operating costs, and reducing NASA's in-house research groups and the teams it used to manage contractors was one path Myers saw towards lower costs. Second, at North American, Myers had worked on Air Force contracts before he became program manager for the Apollo command and service modules. He had experienced some lean Air Force management teams, as well as the army of NASA engineers that had worked on the Apollo project. Myers thought NASA's programs would be best served by staffs somewhere between these extremes.[35]

Finally, the contractors were strengthening headquarters' ability to effect the design it wanted. When given the Air Force-NASA requirement for high cross-range, all of them were coming back with delta-wing configurations. Each, moreover, had different ideas on how the shuttle should be configured. The profusion of ideas, and the engineering studies of them, were giving NASA the data it needed to make the optimal trade-offs among configurations. With these data coming in, and with the conviction that the Faget design would not serve, Myers was able, in January 1971, to remove the DC-3 shuttle from the design options.[36]

Leroy E. Day, who was then deputy director of the shuttle program at NASA headquarters, later recalled, "Dale Myers turned out to be a real champion. . . . For a long time we were still continuing to argue about this low cross range, high cross range or the straight-wing versus the delta-wing configuration. And Dale was the one, I think, that really held the line and said, 'No, we're not going to go for this straight-wing business. We're going to go for a delta-wing vehicle' . . . and by that time, there was lots and lots of evidence from experimentation done within and outside of NASA that the straight-wing configuration had so many limitations that we really ought not to embark on that."[37]

In the struggle between Faget's group at Houston and the Office of Manned Space Flight at headquarters, both sides made use of contracts with industry. Faget used them to add engineer-power to the staff he already had in-house, to deepen the base of data for the DC-3 shuttle, and to obtain independent confirmation of his view that it was a feasible configuration.[38] It is a truism in the history of technology that a more studied concept has a better chance of winning the funding. It will appear more "fully baked," more trustworthy, than less developed but possibly better ideas in the field. Myers used the contracts underway for NASA to show that the delta-wing configuration had wide industry support, and to generate the design ideas and data needed to produce an optimal design.

The use of contracted studies to buttress the position of one or another group in the government is not, of course, unusual. The claim I wish to make here, however, is that the political use of contracted studies functioned as a mechanism for drawing industry ideas into the shuttle design. It became a second factor, alongside the experience industry had acquired through a decade of participation in the study of spacecraft.

THE THIRD factor was the fiscal pressure the Nixon administration put on NASA. At the time that Nixon came to office, the federal budget was showing the then alarming shortfall of twenty-five billion dollars and inflation seemed high. Nixon embarked on a conservative policy of cutting the cost and size of government.[39] The Bureau of the Budget, in 1970 reorganized into the Office of Management and Budget (OMB), became one of his instruments for restraining spending and a principal actor in the attempt to cut back NASA. The OMB group that monitored NASA, moreover, was critical of the manned part of the space program and hence of projects like the shuttle.[40]

To overcome OMB skepticism, NASA in 1970 contracted for a number of external studies on the economics of a shuttle. One contract went to the Aerospace Corporation to study payload and launch costs, a second to Lockheed to see how the use of a shuttle could reduce the costs of building payloads, and a third to Mathematica, an economics consultant firm, to compare the costs of launching spacecraft by shuttle and by expendable launch vehicles for both government and commercial missions. These studies showed that "any economic justification depended crucially on the shuttle 'capturing' all U.S. missions . . . during the 1980s . . . [including] all military and intelligence payloads." This made the Department of Defense more crucial than ever to the fate of the shuttle, because its support was needed for both economic and political reasons.[41] It was a somewhat ironic outcome, because the DoD requirement for high cross range increased the weight, and hence the expense, of the shuttle at the very moment when the OMB was trying to drive down the funding.

Thomas Paine resigned as NASA administrator in July 1970, just as these studies were beginning. In March 1971, Nixon appointed James C. Fletcher to replace him. It had already become clear that NASA could not win funding for both the space station and the shuttle. Indeed, the concept of the shuttle had been quietly changing. Although NASA still intended that it be used to build and resupply an eventual space station, for the immediate future it was being conceived as a kind of short duration station itself, one that could not only release and retrieve satellites but could perform station-like duties as it orbited for seven to thirty days. Fletcher decided to put his energies into the fight for this reconceptualized shuttle.[42]

In May, the Office of Management and Budget told Fletcher that NASA must reconcile itself to administration budget requests for the next five years at not much over three billion dollars. This doomed the fully reusable shuttle that NASA and its contractors had been designing. That shuttle had been projected to cost ten billion dollars to develop, and in peak years, the expenditures would reach two billion dollars. Under the new dispensation, NASA would have to restrict peak spending on the shuttle to one billion dollars if it wished to maintain other programs, with a total development budget of five to six billion. Only a system that was partly expendable could be put together for this price.[43]

Forced to retreat from its original project of a fully reusable shuttle, NASA began to consider one idea after another. Many of them owed as much to industry as to NASA itself. The external expendable tank was brought back as a possible cost-cutting concept. So too were expendable boosters. The agency also mandated a study of "phased development." For the orbiter, this would mean developing two versions, a "Mark 1" and "Mark 2." Mark 1 would have off-the-shelf engines taken from Apollo hardware. Mark 2 would come later and would use the innovative space shuttle main engines that NASA was planning to develop. A two-phase booster program ("Block 1" and "Block 2") would see the first shuttles flying with expendable boosters, and only the later versions would use the fully reusable flyback boosters that Rockwell and McDonnell Douglas were studying on their Phase B contracts.

The mainline Phase B contracts of North American Rockwell and McDonnell Douglas were no longer mainline. The two companies were directed to add work on configurations with expendable tanks to their studies of fully reusable systems. The Alternate Phase A contracts, and especially the Grumman contract, gained in importance. Now the whole panoply of booster options that had been studied under Alternate Phase A were on the table. They "included horizontally landing flyback boosters, both piloted and unmanned, and fully and partially-expendable boosters based on . . . Saturn S-IVB [stage] technology. . . . The strap-on booster options encompassed solid-fueled rocket boosters, . . . liquid-fueled pressure fed ballistically recoverable boosters, . . . and liquid-fueled pump-fed boosters." Both parallel burn, in which the engines of both the booster and orbiter were lit at launch, and series-burn, in which the booster was ignited first, were evaluated. By August 1971, at least ten designs were under serious consideration.[44]

In this extraordinarily fluid situation, aerospace companies tried hard to sell NASA on the technologies they knew best. They hoped—vainly as it turned out—that a shuttle design that incorporated their own hardware would give them a better chance of securing a production contract. Grumman was pleased when the expendable, external tank was brought back into the mainstream. "[E]xpendable orbiter fuel tanks . . . [have] been pushed by Grumman Aerospace Corp. . . . Grumman, which was not chosen as a Phase B participant, is striving to bring its Phase A effort up to where it will be qualified to bid on the Phase C and D work."[45] Grumman was also glad to see phased development on the table, because it had done some early work in this area. "The Grumman-Boeing team recognizes that it has no lock on these concepts. . . . But it believes that the fact that it was the first to put all these ideas together and has studied them longer than the other contractors gives it a leg up in the shuttle competition."[46]

Martin Marietta tried to sell a version of its Titan as a Block 1 booster. United Technology Center president Barney Adelman visited Fletcher to explain that his company's solid boosters could do the job more cheaply. Boeing brought NASA a suggestion for a modification of the S-1C stage the firm had built for the Saturn V rocket that was launching the Apollo. This would be a first phase "expendable" booster for the shuttle, but Boeing offered to design it so that it could actually be reused a handful of times. "[Peter] Downey [Boeing manager] believes the Boeing proposal for Block 1 and Block 2 development is an ideal way to resolve the hard economics faced by the space agency in its push for shuttle dollars."[47]

It was only during the fall and winter of 1971-1972 that NASA rejected phased development, and settled on the shuttle system that is now familiar, with its central external tank and strap-on boosters. It is hard to apportion credit for the pivotal ideas, because NASA and industry teams were working so closely together. Faget's team, working with its Grumman-Boeing contractors, had the crucial insight in the spring of 1971 that an external, expendable tank containing both fuel and oxidizer could serve as the structural spine of the shuttle system. The Lockheed STAR Clipper external tanks had formed a triangle within which the orbiter nestled, and one difficulty with the Lockheed design was the cumbersomeness of shedding this triangle of tanks. Other designs had located two expendable tanks to either side of the orbiter fuselage, or under its wings.[48] Making the tank a structural element transformed it into something more useful than a mere container and it became correspondingly more attractive.

Next, Grumman and McDonnell Douglas, working independently, combined this central external tank with strap-on boosters that would burn in parallel with the orbiter's engines. After that, "MSC Spacecraft Design Division engineers combined these structural insights with the fruits of extended aerodynamic work on delta wing orbiter configurations," and the final design emerged.[49] At the end of the winter of 1972, NASA decided on solid rocket motors rather than liquid propellant engines for the strap-on boosters, on the grounds that they would be easier—hence cheaper—to develop.[50]

Studies of the space shuttle main engine (SSME) proceeded indepen-

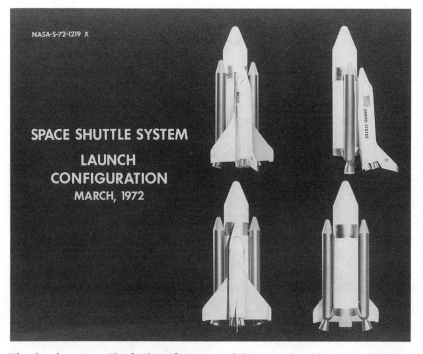

NASA-S-72-1219 X

SPACE SHUTTLE SYSTEM
LAUNCH
CONFIGURATION
MARCH, 1972

The shuttle system. The final configuration of the space shuttle was established in early 1972.
Source: Published in Guilmartin and Mauer, *A Shuttle Chronology,* VI-46. NASA-S-72-1219X, NASA Headquarters History Office, Washington, D.C.

dently. Here too, NASA was very open to industry ideas, and the final design used elements from each of the three firms—Aerojet, Rocketdyne, and Pratt & Whitney—that had had contracts for Phase B studies.[51]

In summary, OMB's insistence that NASA budgets stay constant was another reason that industry had input into the shuttle design. OMB pressure, by forcing a redesign of the shuttle, sent NASA scrambling for new ideas and aerospace industry engineers were able to provide many of them.

IT IS instructive to compare the design of the shuttle with the design of the Mercury and Apollo capsules a decade earlier. Mercury and Apollo had simple, clear-cut missions and were designed on tight schedules. In contrast the shuttle had multiple functions and there was disagreement within NASA as to their relative importance.

There were also institutional pulls. The Apollo program was drawing to a close and the centers needed projects to keep themselves going. The

An alternate external tank configuration. This Grumman-Boeing design from mid-1971, executed under an alternate Phase A shuttle contract, features two external tanks for the liquid hydrogen fuel mounted on the orbiter's fuselage.
Source: NASA Headquarters History Office; originally published in Guilmartin and Mauer, *A Shuttle Chronology,* V-245.

kind of work the shuttle could offer, for example, the chance it could offer Marshall to develop new large engines, became important. On top of this, OMB waged an unremitting fight to keep shuttle development costs low. Rapid design was out of the question and the slower process that replaced it opened room for a multitude of design ideas.

Symbolic of the change was the dismissal of Faget's DC-3 design. In the Mercury program, Faget's capsule design had simply been presented to industry as a fait accompli. In the Apollo program, NASA contracted with three firms for spacecraft design studies: nevertheless, it was clear early on that Faget's design was going to prevail. What a dramatic contrast the DC-3 shuttle presents! The difference provides a striking sign of the transformations NASA, industry, and the nation's political and economic situation had undergone.[52]

The Shuttle Contracts

In January 1972, President Nixon approved the shuttle and in the spring, Congress voted funding for it. Aerospace firms could see some disadvantages to the contracts for shuttle hardware that were now imminent. It was clear to industry that the shuttle would have to be built under tight financial constraints. Moreover, it would likely be continually under attack from Congress and the Office of Management and Budget. Because NASA was counting its pennies, firms would also feel themselves constrained to underbid in order to win. On the other hand, the shuttle was to be an operating transportation system rather than an Apollo-like one-time spectacular. The firms that got the contracts might therefore look forward to orders for a fleet of shuttles, a fleet that in 1972 was being projected as between five and nine.[53] Beyond that fleet could come upgrades and second-generation craft.

The shuttle also differed from Apollo in that the orbiter was the world's first cut at a passenger-carrying space plane. Some of the innovative technologies created for the shuttle might be transferable to military or commercial aircraft. Still more important, no one knew how significant space planes themselves might become. This, therefore, was technology to which companies that manufactured aircraft wanted access. They were not likely to get that access through military contracts. The Air Force's Manned Orbiting Laboratory—intended as a manned reconnaissance satellite—was canceled by the Nixon administration in 1969. The concept had been questionable, the costs had mounted, and new unmanned reconnaissance satellites had preempted the mission. After the cancellation, the military drew back from manned space, leaving the field to NASA.[54] And the shuttle was clearly going to be NASA's only manned space initiative of the 1970s. To keep a hand in, aerospace companies needed to work on it.

Perhaps the most compelling reason for interest in the shuttle contracts, however, was the depression in which the aerospace industry found itself. U.S. government orders for military aircraft and space hardware had been declining, and commercial orders, instead of acting to counter the decline, had gone down as well. Widespread antagonism toward military spending put additional pressure on defense firms.

Military procurement started downward in 1969. From a fiscal year 1968 high of forty-four billion dollars, it would reach a low, in constant 1978 dollars, of seven billion dollars in 1975. Decreasing departmental budgets, in contract dollars, were partly to blame, but so was a reorganization that put a higher percentage of DoD's money into personnel and research and development. In aircraft, the drop was from seventeen billion dollars in procurement in fiscal year 1968 to seven billion in 1975, in 1977 dollars.[55]

At the same time, the progression to ever more complex and expensive weapons systems continued. A state-of-the-art military plane could be bought for a few million in the 1950s; in the 1970s it cost twenty million or even fifty million dollars. As a result, the Pentagon was putting its procurement dollars into fewer contracts. This meant that only a handful of new contracts were being awarded; each contest became more of a make or break situation for the competing companies.[56]

In the nation at large, the struggles against Nixon, who was extending the Vietnam war to Laos and Cambodia, fanned a generalized revulsion against war and the military. Everywhere, there were calls to divert federal dollars from defense to areas like housing, the environment, and alternate energy sources. As this antimilitary mood took hold, Congress became more intolerant of cost overruns on weapons system contracts. This, in turn, put pressure on companies with defense contracts, a pressure the more ironic because the increasing complexity of the new weapons was making accurate estimation of their costs more difficult.[57]

Sales of military planes to foreign governments now became an increasingly important part of aerospace industry revenues. But here also the prospects were clouded by the growing competition from Europe. The United Kingdom and France had rebuilt their aircraft industry. West Germany was beginning to do so, and Italy was waiting in the wings. Already during the 1960s, U.S. share of foreign military aircraft had fallen from 97 percent in 1962 to 68 percent in 1968.[58]

The market for commercial passenger transports, meanwhile, was glutted with offerings of new planes. Boeing was developing the 747, McDonnell Douglas the DC-10, and Lockheed, which had not been in this market since the failure of its Electra in the early 1960s, was trying its luck again with the L-1011, a three-engine jet. Airline passenger traffic had grown rapidly during the 1960s, averaging an increase of 17 percent a year. The airline industry, which was federally regulated, had competed on the basis of service and the quality of aircraft rather than prices and had, therefore, regularly updated its fleets with state-of-the-art planes. But growth in passenger traffic leveled off toward the end of the 1960s. Airlines began to be plagued by overcapacity, high costs, and low earnings, and they cut back on their orders for new planes. Boeing, McDonnell Douglas, and Lockheed were all having difficulty lining up enough orders to break even, where income would equal development costs. In this market also, moreover, the U.S. companies faced the prospect of European competition.[59]

While NASA's expenditure for space declined after 1965, military space budgets went down even faster. The cancellation of the Air Force's Manned Orbiting Laboratory eliminated one large military program. Advances in satellite technology made for lower spending because the defense and intelligence communities could field longer-lived and better-performing satellites, and thus cut back on the number of satellites and launches needed.[60]

Declining orders exacerbated the strains that already existed within some of the aerospace firms. In the recently conglomerated North American Rockwell, they aggravated the tension between the North American Aviation faction, headed by company president Atwood, which wanted to pursue government contracts, and the Rockwell faction, headed by chairman and CEO Willard Rockwell, which sought to shift more of the company's effort into commercial products. When North American Rockwell lost the competition for the F-15 fighter plane to McDonnell Douglas at the end of 1969, Atwood retired, and Rockwell brought in as the new president Robert Anderson, who had come to North American Rockwell in 1968 after years at Chrysler.[61]

In 1970, North American Rockwell won a contract for research, development, and prototypes for the B-1 bomber. Were it to go into production, it would replace the B-52 as the main Air Force aircraft of that type, and contracts worth up to fifteen billion dollars could be in store. But President Nixon was seeking a détente with the Soviet Union, and it was impossible to estimate how many B-1s would actually be ordered from the company. Willard Rockwell and Robert Anderson continued to acquire commercial firms. By September 1972, 60 percent of North American Rockwell sales were in the nonaerospace groups, up from less than 30 percent at the time of the 1967 merger. Shortly after, Rockwell absorbed into North American Rockwell his own Rockwell Manufacturing Company, and in early 1973, he changed the company's name to Rockwell International Corporation.[62] The name "North American," known and respected in air and space for forty years, disappeared.

Other companies were also troubled. General Dynamics, the defense conglomerate that was Convair's parent company, had 83 percent of its sales with the government in 1969. It was taken over by a major stockholder, Henry Crown, who brought in a new president and tried to shift the company toward commercial products.[63]

Lockheed, one of the country's largest firms, and like General Dynamics, almost entirely in the defense business, was in the worst shape of all. In 1965 it had underbid on and won, the C-5A, a military transport, and was now finding that a Congress steeped in the national antagonism to the military was in no mood to forgive units cost over-runs. On top of this, Lockheed's contract for an advanced Army helicopter, the Cheyenne, was canceled. It had borrowed hundreds of millions of dollars to finance the L-1011 passenger jet for which it was having such difficulty lining up customers. In early 1971, Rolls Royce, the British company that was making the L-1011 engines, declared bankruptcy because it was spending more on the engine's development than Lockheed was able to pay. Lockheed went hat-in-hand to the Congress for a $250 million loan guarantee to forestall bankruptcy.[64]

Not every aerospace firm was doing badly. Hughes Aircraft Company increased its sales, year by year. TRW was well diversified, and Martin Marietta was on solid footing, except for its unwise acquisition of an aluminum company in 1969.[65] Nevertheless, the industry as a whole badly needed the four-plus billion of new business that NASA's shuttle could bring.

THE REQUEST for proposals for the first shuttle subsystem, the main engine, was issued in March 1971, as engines are the part of air and spacecraft that require the longest development time. It was to be a liquid-propellant rocket engine of advanced design. At that point, it was not at all clear how much the contract would be worth. Fully reusable boosters with engines that permitted them to return to earth on their own power were still under consideration. If used, they would be outfitted with variants of the same main engines that would power the orbiters. It was therefore possible that of the order of fourteen engines might be used for each shuttle: twelve, for example, on the booster, and two on the orbiter. A modest fleet of seven shuttles would then require the production of about a hundred engines, and the contract might be worth as much as a billion dollars.[66]

A contentious fight among bidders delayed the final contract until the following year. Three rocket engine companies had performed prior studies on the shuttle engine for NASA and were asked to bid on the project: Aerojet General, North American Rockwell's Rocketdyne Division, and United Aircraft's Pratt & Whitney Division. In July 1971, NASA awarded the contract to Rocketdyne. Pratt & Whitney immediately filed a protest with the General Accounting Office (GAO), arguing that NASA's decision ignored eleven years of solid experience on the XLR 129 engine Pratt & Whitney built for the Air Force in favor of a mere "paper" design of Rocketdyne's. The GAO rendered its verdict, upholding NASA's decision, in the spring of 1972. By that time, the shuttle's configuration had been fixed. The flyback booster had been eliminated and the only main engines were to be those on the orbiter, which was to be fitted with three each. Rocketdyne's initial contract called for twenty-five engines, at a total price of $450 million.[67]

The contest for the orbiter was less litigious, although more was at stake. This was to be the largest shuttle contract: in January 1972, NASA estimated it would total over three billion dollars through 1980.[68] Four companies sent in proposals in response to a request sent out in March 1972: McDonnell Douglas and North American Rockwell, which were carrying out Phase B work for Marshall Space Flight Center and Houston's Manned Spacecraft Center, respectively, Grumman, which, with Boeing, was doing an Alternate Phase A study for Houston, and Lockheed, which had an Alternate Phase A contract with Marshall. NASA rated Grumman and North American Rockwell the top candidates, with Grumman first on the technical side and North American Rockwell first on management. NASA choose Rockwell, reflecting the importance of management competence to the agency.[69]

So far, both shuttle contracts had gone to Rockwell, although NASA was sensitive to the need to distribute contracts over many companies and geographic areas.[70] Immediately after the orbiter award, NASA took steps to ensure that at least some of the money would be passed through to the losers in the form of subcontracts. George Low, who had moved up from Houston to become NASA deputy administrator in December 1969, wrote to a White House staff member on August 14, 1972, "Our plan, which we will start implementing this week, calls for the following: 1. We will identify positive hardware elements for eventual subcontracting by Grumman and McDonnell Douglas. 2. We will specify home plant locations of Bethpage and St. Louis. . . . 4. North American will be asked to respond with a plan by August 18. . . . Our procurement regulations permit this kind of sole source subcontracting."[71] To the press, Administrator James Fletcher said "If I were the prime contractor, I'd look at such companies very closely because of the expertise they have acquired on the shuttle project."[72] Grumman got the wing of the orbiter, General Dynamics-Convair the mid-fuselage, McDonnell Douglas the system that allowed the orbiter to maneuver while in orbit, and Fairchild Industries the vertical tail fin.[73]

The last two major contracts, for the external tank and the boosters, served to broaden the circle of contractors still further. The external tank contract went to Martin Marietta in August 1973. According to the company's official historian, Martin Marietta had not even planned to bid. It was NASA Administrator Fletcher who encouraged the president of the Aerospace Division to "go after it."[74]

The solid rocket motors were awarded to the Utah-based Thiokol

Chemical Corporation in November 1973. It was a controversial decision, as Fletcher, a Utah Mormon, had been under heavy pressure from Utah notables and the state's Congressional delegation to give the contract to Thiokol.[75]

NASA also kept Lockheed in the program as supplier of the ceramic material that was to protect the shuttle from reentry heating. The decision was based on a "shoot-out" held in a Johnson Space Center test oven among Lockheed's ceramics and competing materials from General Electric, Martin Marietta, and McDonnell Douglas. When the oven was opened after the heat trial, only the Lockheed compounds remained intact. Lockheed engineers believed that NASA also chose Lockheed because it wanted Lockheed expertise and the Lockheed culture as part of the shuttle team.[76]

One student of policy has written that by the end of the contracting process, "every major space technology firm [got] a juicy contract."[77] This is probably an overstatement. Northrop did not compete. Boeing did but got nothing, although a Boeing 747 was purchased from an airline to ferry the shuttle from Rockwell's California factory to the launch pad. Nevertheless, much of the industry did get contracts. On the one hand, this reflects the fight that the companies mounted. The no-holds-barred rivalry of aerospace firms has been described by others.[78] In the early 1970s, it was made still fiercer by a situation of gross underutilization of capacity. Accordingly, the firms fought with every weapon: low bids, appeals to Congress, litigation, advertisements, and the lobbying of NASA officials. On the other hand the evidence, and particularly the subcontracting, also suggests that NASA wanted to spread employment across firms and across regions, and thereby to ingratiate itself with Congress, the administration, and the industry.

The Shuttle Transformed

There was a profound difference between building a shuttle to service a space station as part of the saga of human exploration of space, and building a shuttle that had no space station to go to. The shuttle's role had changed over the half-decade from 1969 to 1974. NASA had come to stress the economic payoff: the shuttle would be a cost-effective system for placing in space (or retrieving or servicing) military, scientific, and commercial satellites.[79] It was a metamorphosis with repercussions for the aerospace industry in areas ranging from rockets and satellites to prospects for manufacturing in space.

In rockets, the salient fact is that a space transportation system for emplacing satellites was already in existence. This was the system of expendable launch vehicles, including Martin Marietta's Titan, McDonnell Douglas's Delta, and General Dynamics' Atlas. The intention of NASA was that the shuttle take over most of the missions of expendable rockets—that is to say, take over the markets of McDonnell Douglas, General Dynamics, and Martin Marietta. These companies had all received shuttle contracts, even though North American Rockwell won the lion's share. The companies, therefore, were not being squeezed out of space transportation. Rather, the shuttle promised to shift the market shares held by the various aerospace companies.

In the area of manufacturing in space, NASA's desire to fly the shuttle as frequently as possible was key. A large number of flights—provided of course, that the cargo bay was full on each of them—would allow the agency to amortize the cost of the shuttle over many missions and thus justify its claim that the shuttle would make access to space cheaper. NASA therefore proposed not only to launch, retrieve, and service spacecraft with the shuttle, but also to use it to fly some cargoes in a "sortie" mode, where the equipment brought up would remain with the shuttle, orbiting with it during its days aloft, and returning with it when it landed. One candidate for shuttle sorties was science or technology experiments. Another was experiments aimed at the fabrication of commercial products in space.

NASA's interest in space manufacturing had the effect of inducing a like interest in aerospace companies. The companies, after all, also had a stake in the shuttle's success. They were, moreover, eager to see the shuttle program followed by a space station program. And, even more than the shuttle, a space station would find justification in its usefulness as a site for the processing of products. Finally, the companies were interested in pleasing NASA and in demonstrating expertise in those areas NASA cared about. Thus, NASA's need to maximize traffic on the shuttle drove the aerospace industry to a livelier appreciation of the industrialization of space.

For satellite makers and satellite owners, the shuttle's new role as a space transportation system brought uncertainty. The manufacturers had designed their satellites to mate with expendable rockets. Were they to shift to the shuttle, they would need to redesign these craft for a man-rated vehicle. NASA was promising all kinds of advantages to a shuttle launch, not the least of them price.[80] At the same time, the shuttle was a new technology. No one knew for certain whether it would work, how reliable it would be, or how much it would cost to operate.

AT THE time the shuttle contracts were signed, the United States was the only noncommunist nation launching satellites, and NASA and DoD were the only U.S. organizations authorized to carry out the launches. NASA launched the Scout, Atlas-Centaur, and Delta rockets using facilities rented from the Air Force at Cape Canaveral, Florida, and Vandenberg Air Force Base, California. The Defense Department launched the Thor-Able-Star, the Agena, the Atlas, and the Titan.

To carry out its launches, NASA bought rocket stages and launch services from aerospace companies and funded the companies to conduct rocket upgrades. It bought the Scout from Vought Corporation for small payloads, the Delta from McDonnell Douglas Aerospace Company for mid-sized loads, and the Centaur from General Dynamics to combine with the Air Force's Atlas into a two-stage booster for large payloads.[81] The payloads it sent up with these rockets fell into three categories. Part of them were NASA experiments and planetary probes. A second group were spacecraft launched for other government agencies. Finally, NASA launched satellites on a cost-reimbursable basis for foreign governments, for multinational consortia like Intelsat, and for private firms.

Industry, therefore, played the role of a contractor to NASA, supplying the agency with hardware, launch services, and design services. But each company also had an obvious stake in maximizing the number of launches made with its own rockets. The greater the number of launches, the more the hardware that NASA would buy. Thus, although ostensibly it was NASA that marketed launches, in fact, company sales teams worked hand-in-hand with NASA program managers in finding customers. McDonnell Douglas, for example, had marketing groups closely following satellite activity world-wide and trying to line up each mid-sized satellite, as it came on line, for the NASA Delta program.[82]

The mere fact that NASA had chosen the Delta for all its mid-sized payloads did not remove McDonnell Douglas from competitive pressures. The Air Force vied with NASA for control of weather satellite launches in the early 1960s. A little later, it competed with NASA for launches of communications satellites.[83] To the extent that these competitions used rockets made by different companies (for example, it was Lockheed's Agena rocket that the Air Force was pitting against McDonnell Douglas' Delta in the area of communications satellites) the competition between NASA and the Air Force became a proxy for competition among companies. Thus, although in a formal sense there was no U.S. commercial launch industry in the early 1970s, under the surface there was marketing by private firms and competition for market share among them.

This quasi-commercial enterprise, with its de facto management by NASA-industry teams, was slated to come to an end once the shuttle began operation. Expendable launch vehicle manufacturers were thus faced with the following prospect. Until the shuttle started operation at the end of the decade, their rockets would continue to be purchased. Contracts for upgrades and improvements, however, would be severely limited. Once the shuttle started flying, they would enter a brief transition period during which they could expect to sell a few expendables as back-ups to cushion users against shuttle delays or malfunctions.[84] And after that, there would be no further market for these rockets.

One option open to the companies was to try to influence federal decisions on space transportation. Despite all the agreements of the early 1970s, there was still plenty of debate over the proper role of the shuttle vis-à-vis expendable launch vehicles. The Department of Defense had signed on with the shuttle but was lukewarm in its support. There were shuttle opponents in Congress and the scientific community. In 1975, the General Accounting Office of the U.S. Congress openly advocated that the shuttle be treated as a research program and that expendable rockets be maintained and upgraded.[85] The companies' clout included not only their own connections with Congress, NASA, and the Defense Department, but also the pressure they could bring to bear through their subcontractors.

But there were other options, among them the possibility of developing product lines that supported the shuttle. A company could adapt its rocket expertise, for example, to the design of an "upper stage." Because it was to travel in a low earth orbit, the shuttle would not by itself be capable of placing a satellite in a higher trajectory such as a geosynchronous orbit; some sort of auxiliary or "upper stage" rocket would be needed. Originally, NASA planned to develop the Space Tug for this work. The Space Tug was to have been a reusable system, one that could not only place satellites into and retrieve them from geosynchronous orbit but would carry humans there and back. But NASA budgets were too cramped for a Space Tug to be fitted in during the 1970s and the agency postponed this project. Instead, the Air Force undertook to finance an interim upper stage (IUS). The rocket makers were very interested in the IUS. General Dynamics offered its Centaur, McDonnell Douglas its Delta, and Lockheed its Agena, while Martin Marietta and Boeing had candidates they called Transtage and Burner 2, respectively.[86]

In 1976, the Air Force chose Boeing, but this did not end company interest. NASA thought that other kinds of upper stages might be needed in addition to the IUS, and suggested that companies might try to develop them for sale to satellite makers.[87] McDonnell Douglas used its own funds to develop the payload assist module-D (PAM-D) that would function as an upper stage for either the Delta rocket or the shuttle during the transition period when cargoes were being shifted from expendables to the shuttle. It also financed a PAM-A for shuttle launch only.[88] General Dynamics sought to find a place on the shuttle for its Centaur upper stage, then used with the Atlas.[89] Designing and finding a niche for such rockets remained an enterprise for aerospace companies throughout the 1970s and 1980s. Thus, the struggle for market shares in space transportation was not confined to a fight over expendable vehicles versus the shuttle. The size and diversity of the companies allowed them to assay a move into other spaces within the market, and it was not at all evident during the decade what the total pie would be and how it would be divided.

IF THE market for space transportation was ill-defined in the mid-1970s, the situation for space manufacturing or for other activities then lumped under the rubric of "space industrialization" was more uncertain still. NASA had been studying the behavior of materials in space since its inception in the 1950s. It needed to know how liquid propellants flow and combustion transpires in near-zero gravity to improve rocket propulsion and prevent fires. It needed to know how substances melt in space in order to understand welding and soldering procedures for the construction of a space station in low earth orbit. It was also interested in whether space stations might serve as laboratories for fundamental research on the properties of materials and whether specimens might be created in space for scientific study, or even sale, on earth. Experiments to these ends were carried out on three moon trips, *Apollo-14* (winter 1971), *Apollo-16* (spring 1972), and *Apollo-17* (late fall 1972), on the *Skylab* (1973), and on the Apollo-Soyuz rendezvous (summer 1975).[90]

As the shuttle took over the role of NASA's flagship program, NASA expanded its exploration of the processing of materials in space and reoriented that exploration to emphasize the use of space for industry. It commissioned General Electric's Space Division to involve nonaerospace companies in a survey that would identify profitable space-made products and space-based enterprises.[91] Of the General Electric (GE) study, *Aviation Week & Space Technology* explained, "NASA understands that a major problem in exploiting shuttle capabilities lies in a critical missing element—finding paying users for the system in sufficient numbers to utilize this new national resource economically. . . . [B]ut there has been a notable lack of response from the non-aerospace industry, which could become the preponderant customer population of shuttle users and potentially the source of the bulk of income for the program. To tackle this problem, NASA selected GE to do a study."[92]

The roll-out of the first shuttle mock-up, the *Enterprise*, in September 1976 provided an occasion to step up the pitch. Representative Olin Teague took the opportunity to link the event with the dawn of the industrial use of space, and Senator Barry Goldwater predicted that the shuttle program would "loom . . . as one of the biggest bargains this country ever bought," dwarfing even the purchase of Alaska.[93]

The *Enterprise*'s first "flight," a five-minute glide from a height of 22,000 feet on August 12, 1977, was made without the space shuttle main engines and the thermal protection system, the two shuttle components that were beginning to give the most difficulty. Nevertheless, it was another stimulus for hyperbole. *Business Week* told its readers, "The shuttle will open up space to vast numbers of new users. Materials that are prohibitively expensive or impossible on earth will be manufactured in the zero-gravity environment of space. Other applications will at last take advantage of the inexhaustible supply of solar energy in space. And there will be no pollution problems."[94]

As NASA headquarters sought to encourage industry to think about space enterprise, Marshall Space Flight Center also took steps in that direction. The center had had a Space Sciences Laboratory, with specialists in the processing of materials in space, early on. In the 1970s, with the Apollo era over and the NASA centers struggling to define new roles for themselves, that expertise seemed an attractive asset. Marshall hoped it might become the premier NASA center for shuttle utilization.[95]

In 1976, headquarters assigned Marshall to support it in expanding activity in space processing. To formalize its position, the center drew up a charter that listed both research and industrial outreach as its tasks. "Space Processing of materials to obtain highly valuable commercial products for use on earth is rapidly emerging as a major new application of space. . . . The challenge to NASA is to demonstrate the economic advantages of the technology and to encourage industrial participation at an early stage of development."[96]

It was the Program Development Group at Marshall that took on the job of proselytizing industrialists and would-be industrialists. It worked to bring knowledge of the results of the center's Space Sciences Laboratory and other academic and industry research to the private sector. It touted the shuttle as a test-bed for industrial processes. It even helped bring new companies into being. In the late 1970s, it introduced Richard L. Randolph, a retired military officer with an interest in space and experience in finance, to Harry C. Gatos, an MIT professor who had been experimenting with space-borne crystal growing through NASA, and Russell Ramsland, a wealthy Harvard Business School student from Texas. The result was that Randolph and Ramsland founded Microgravity Research Associates (MRA) in 1979, with the proposition of making gallium arsenide crystals in space by a method developed by Gatos and patented by MIT. In 1979, MRA proposed to NASA, and it April 1983 it signed with the agency, an agreement permitting MRA to fly more than a half dozen free experiments on the shuttle, in return for providing NASA with data and promising to follow the flights with commercial production.[97]

The aerospace companies had participated in NASA's space processing projects from the start, but as contractors. Boeing, Grumman, Lockheed, TRW, Rockwell—all had scientists working on the experiments that NASA had flown or on the ground-based research that had served as backup. But as the shuttle took center stage, the companies began initiatives of their own.

Boeing started a small, internally funded program on solar power satellites. The idea had been proposed in 1968 by Peter E. Glaser, then head of Engineering Sciences at Arthur D. Little, a consulting firm. Glaser pointed out that solar power on earth is intermittent because of cloud cover and necessitates large areas of collectors because the sun's energy strikes the earth with low intensity. He suggested collecting solar energy with satellites in geosynchronous orbits where they would receive sunlight nearly 100 percent of the time. The energy could be beamed via microwaves or lasers to a ground station below.[98] The power outages, oil embargo, and increasing concern for both energy and the environment that marked the early 1970s motivated Boeing to take a look at Glaser's ideas.[99] If they were feasible, it would mean a large new market for makers of spacecraft and launchers. It would also push Boeing to develop a capability in space-based construction, for the satellites would have to be built in orbit. This was a field of technology well worth entering, because it could be called on for a variety of tasks, including construction of the space station. In 1976, Boeing became one of two companies (the other was Rockwell) to receive NASA contracts for a detailed design of a solar power satellite.[100]

Rockwell, with the largest stake in the shuttle, set up a Space Transportation System User Service Center, and began to run a series of gee-whiz advertisements on the new possibilities opened up by the shuttle.[101] Grumman embarked on research into crystal growth in space.

The company that committed itself most deeply was McDonnell

Douglas. McDonnell Douglas thought of itself as a pioneer among space companies. It was the first U.S. corporation to build a manned spacecraft; the Mercury capsules were developed and constructed at its St. Louis plant. McDonnell Douglas management wanted its company to remain at the forefront in space. Doing so appealed to its sense of company tradition; it was a matter of company pride;[102] and it protected the image the company had put before the public since the late 1950s.

McDonnell Douglas was, moreover, the most involved in space station work of all the aerospace companies. It had been the main contractor for the Air Force Manned Orbiting Laboratory before that project was canceled in 1969. Together with North American Rockwell, it had been one of the two companies chosen for space station study contracts in 1969 through 1971. McDonnell Douglas had worked on *Skylab*. It had provided technical assistance to the European companies building the *Space Lab*, a module for the belly of the shuttle that was being designed and constructed under the auspices of the European Space Agency, and that was to transport science and technology experiments for the shuttle's sortie missions. It was McDonnell Douglas that in 1974-1975 had performed a paper study for NASA of a conservatively engineered space station under the Manned Orbital Systems Concept contract, and it was McDonnell Douglas that, along with Grumman Aerospace Company, had been charged in 1976 with carrying out studies for a more ambitious station.[103]

McDonnell Douglas believed that a station would be the next large space program to be authorized and it hoped to have a major role in building it. Materials processing in space, though it was now being discussed for the shuttle, was far more compatible with a station. Only the most limited amounts of products could ever be processed during the week or two duration the shuttles would be in orbit. A station, by contrast, would be aloft indefinitely. For McDonnell Douglas strategists, materials processing in space was a way of staying at the frontiers of space endeavor, and also a way of showing NASA the company's commitment to the space station and space industrialization.[104]

McDonnell Douglas had already done some contract work for Marshall Space Flight Center on the range of possibilities for industrial activities in space. It had also done some in-house studies in anticipation of a NASA program to purchase hardware for research on space processing.[105] In 1975, the firm started a program on electrophoresis in space. Electrophoresis is a process for separating and purifying biological materials like proteins, viruses, and cells. On earth, it was only employed in research, because effects caused by gravity limited electrophoretic chambers to diameters far too small to serve industrial applications. In the zero-gravity environment of space, however, separation of biologicals in commercial quantities might be possible. The idea was that McDonnell Douglas would make itself into the designer, developer, and operator of equipment for electrophoresis in space, and it would recruit pharmaceutical houses as partners to identify marketable products, provide the raw materials, and conduct the clinical tests and negotiate for Federal Drug Administration approval.

The program's leaders made searching for a partner a top priority, and in 1978 they concluded an agreement with Ortho Pharmaceuticals Corporation, a subsidiary of Johnson & Johnson. Ortho identified, as a first product for research, erythropoietin, a hormone produced by kidney cells that could be used in the treatment of anemia.

NASA gave the companies a full measure of cooperation. In 1980 it signed with them the first Joint Endeavor Agreement. The Joint Endeavor Agreement was a mechanism that had just been put in place for cost-sharing between the federal government and private firms. No money would change hands, but NASA would offer free launches for industrial processes that were in the experimental stage, while the companies would bear the cost of the equipment and related ground-based research. In exchange for the free launches NASA got some of the in-space operation of these processes for NASA's own research.

At the same time, NASA made use of the McDonnell Douglas work in arguing for shuttle support before Congress. New medicines have always had allure, and, perhaps because members of Congress are older than average, they are especially appealing on Capitol Hill.[106] NASA did not hesitate to fill in its oversight committees on the thousands of lives that could be saved with shuttle and space station-derived medicines.

Many in the scientific community and in industries outside of aerospace were skeptical. In 1978, a committee of the National Research Council published a scathing critique of the materials-processing-in-space program. It castigated the sloppiness of many of the experiments done to that date. It pointed out that work done on manned craft like the shuttle or space stations could be expected to be inferior to work done on free-flying platforms or even on the ground because the shuttle or station crews would necessarily disturb the gravitational fields at the site of the experiments. It saw no evident prospect in manufacturing goods in space, and only limited advantages to doing scientific experiments there.[107]

A former director of the Materials Research Center at the then Allied Chemical Corporation later recalled: "I knew that compared to the binding forces in liquids and gases, gravitational forces are . . . very small . . .

Revolutions are almost always associated with first-order effects . . . Furthermore . . . I knew that transportation costs are usually a minor fraction of the selling prices of products; not a totally dominant fraction . . . This [and other considerations] led me to great skepticism regarding the revenues that the new products might generate. I felt that the ratio of them to normal commercial revenues was likely to be equal to, or less than, the ratio of gravitational to electrostatic forces."[108] NASA and the aerospace industry took note of such criticisms, but continued to be bullish on space industrialization.

"Are you being had or are you using us?" inquired Jim Lloyd, California Congressman, of aerospace executives at the September 1977 "Space Industrialization" hearing.[109] He was asking how much the government should pay for research into high-risk, but potentially highly-profitable, space-based manufacturing processes and how much private industry should put into the pot. The question, however, can also be taken in another sense. How much was NASA using the private companies' materials processing projects to sustain the shuttle and argue for a space station? How much were the private companies using NASA's political needs to mount projects that could bring them high-pay-offs and how much were they simply protecting their shuttle contracts and positioning themselves for the competition to come for space station contracts? "One hand washes the other," says the old proverb. But the area of space-based manufactory also shows how thin the line can be between industry as a contractor to government and industry as a commercializer of new technology.

IF NASA encouraged research into space manufacturing with an eye toward providing the shuttle and space station with customers in the long term, it sought to book communications satellites on the shuttle as a way of ensuring customers for the short term. In the years that the shuttle was being designed and built, the communications satellite business was evolving from a monopoly into a lively and fiercely competitive industry. In 1969, there was a single operator of satellite communications in the Western world. This was Intelsat, the international consortium inaugurated in 1964, which provided transoceanic service with satellites made by the U.S. firms of Hughes Aircraft Company and TRW. The Soviet world had Molniya.

When new technology made it possible to focus satellite antennas on smaller terrestrial areas in the late 1960s, some governments opted to operate domestic communications systems. Canada inaugurated a system and Indonesia began planning a satellite system to cover its many islands. In the United States, the Nixon administration opened the U.S. domestic satellite business to competing private companies, and a half dozen firms or consortia of firms applied for operating licenses. The growing number of satellites being launched or planned meant a growing market for launches. It was this market that NASA wanted for the shuttle.

NASA saw sound reasons for satellite operators to book on the shuttle. First, shuttle launches would be priced lower. In the early 1970s, cost estimates were wildly optimistic, running as low as one tenth the cost, per pound of payload, of Delta launches.[110] Second, a NASA-commissioned Lockheed study projected that the shuttle would make it possible to build cheaper satellites. The shuttle cargo bay would be much more capacious than the payload structures of the expendable rockets and the payloads it could carry would be five to ten times heavier. The Lockheed team concluded from this that satellite makers would no longer have to spend extra money to make their spacecraft compact and lightweight. Money would also be saved, they claimed, because satellites would no longer need to be made with folded solar panels or the machinery to deploy them after launch. With the greater space available in the shuttle cargo bay, solar panels could be deployed from the start and hence bolted rigidly and cheaply to the satellite. Still more significant from the viewpoint of cost savings, according to this team, was the fact that the shuttle, as a ferry designed to move back and forth from earth to orbit, could provide for the retrieval or repair-in-orbit of satellites. Lockheed recommended that satellites be given a modular construction. Then the parts that wore out or the fuel that was used up could be replaced in orbit. Other modules could be updated yearly or biannually.[111]

Had satellite makers in fact set out to rework their craft to make use of the shuttle's capacities to haul larger and heavier payloads, or to retrieve satellites, the shuttle, or shuttle-like vehicles, would, of necessity, have captured all traffic. And, in fact, the preeminent satellite manufacture of the early 1970s, Hughes Aircraft Company, did work up one satellite for military use so wide that it could only be launched by the shuttle.[112] The idea of making modular satellites that were refurbishable in orbit would have foundered, however, if for no other reason than the fact that the communications satellites were being placed in geosynchronous orbits 22,300 miles above the earth, while the shuttle would fly in an orbit only 200 miles high.

But there were reasons to reject the construction of refurbishable satellites other than the mere fact that nobody could reach them to do the refurbishing. Modifying satellites to make them modular—with easily replaceable subunits—would bear a significant cost in terms of engineering effort.

Returning them to earth for refurbishing would mean the extra cost of launching them a second time. Moreover, satellite technology was changing so rapidly that it made little sense to extend satellite life beyond the seven to ten years it then measured.[113] In addition, it was important to the owners of satellites that wherever possible their craft be capable of launch on at least two different kinds of boosters. This would give them a backup, in case one type of launcher failed. It would also allow them to take advantage of any future price competition. Hence, the idea of refurbishable, modular satellites was doomed from the start.

Further, and most important, expendable launch vehicles placed satellites destined for geosynchronous orbit directly into transfer orbits—extremely elliptical orbits whose perigee (closest approach to the earth) was about 125 miles but whose apogee (farthest distance from the earth) was 22,300 miles. The satellites had built-in apogee kick motors that converted their orbit into a circle at apogee. The shuttle, however, only placed satellites into low-earth circular orbits. A perigee kick motor had to be provided to move it into the transfer orbit . . . a motor that added several million dollars to the cost and an extra weight that could be as much as the original satellite and its apogee kick motor together.[114]

What did attract satellite owners was the shuttle's low price. Launch prices represented roughly as much of the cost of putting up a communications system as the satellites themselves. In the highly competitive world that the satellite business was becoming, costs were of vital interest. NASA continued to project low launch fees. In 1977, it issued a schedule of tariffs for the first three years of shuttle operation that specified the price for a half-ton satellite as roughly eleven million dollars.[115] Meanwhile, costs for Delta and Atlas Centaur launches were escalating. An Atlas Centaur launch, priced at fourteen million dollars in 1970, would be forty million in 1980. A Delta launch, about ten million dollars in 1974, was on its way to a 1980 figure of twenty-six million dollars.[116] The high inflation of the 1970s, especially the spurt in fuel prices, was one culprit for the rise in expendable launch vehicle prices. But another was NASA itself, which was removing some of the covert subsidies with which it had supported Delta and Atlas Centaur launches, such as low or zero charges on NASA launch and storage facilities.[117]

Satellite owners were also enticed by the shuttle's promise of reliability. This seems prima facie paradoxical. The shuttle was a radically new technology and the technical and business communities might rather have been expected to treat it warily. But the shuttle was to be operated by humans, and the assumption was widely shared that NASA would spend whatever it took to protect the astronauts. This assumption translated into lower insurance premiums, another business consideration satellite owners kept an eye on.[118]

Armed with low prices and expectations of reliability, John F. Yardley, NASA associate administrator for space flight, and Chester M. Lee, director of the Office of Space Transportation Systems, went out to sell the shuttle to communications satellite operators. A full manifest for forthcoming flights would have one benefit among others; it would help convince the incoming Carter administration that industry valued the shuttle as a launch vehicle. Intelsat's new generation of satellites, the seven Intelsat Vs that the consortium was then ordering from Ford Aerospace Corporation, (a subsidiary of Ford Motor Company), was a primary target. Intelsat's board of governors had intended the Intelsat V to be launched on Atlas Centaurs. But NASA approached them with a carrot and a stick. The carrot was an estimated price three fifths the price of an Atlas Centaur launch. The stick was NASA's determination to close down its Atlas Centaur launch facilities, once the shuttle became fully operational.[119]

By 1977, Arianespace was also starting to line up customers for the Ariane rocket, scheduled to come on line in synchrony with the shuttle. Ariane was also likely to boast government-subsidized prices. NASA, moreover, was handicapped in its marketing efforts by enabling legislation that had established it as a research and development organization and not as a commercial operation. NASA representatives, for example, had no expense accounts for taking potential customers to dinner. They had no authority to meet demands that foreign governments were making for offsets.[120] NASA scrambled for ad hoc arrangements to get around such drawbacks. And it relied on industry allies like McDonnell Douglas, whose teams, out selling their Payload Assist Module upper stage, also functioned as salesmen for the shuttle.[121]

Through the summer of 1978, NASA projected that the shuttle would have completed its test flights and begun its commercial flights in mid-1980. The agency formulated rules for the transition from expendable launch vehicles to the shuttle for its commercial customers. It stipulated that expendable launchers would be available through NASA even after the shuttle flew, but only for a few years and at increasing prices. Communications satellite executives were handed a labyrinthine schedule of decision points. During the changeover, they were to choose among shuttle, expendable launcher, and shuttle-with-launcher-for-back-up, with each decision bearing a cost that differed according to the date on which the decision was made.[122]

Meanwhile, the predictable was happening; the shuttle schedule was slipping. Some of the components were taking longer to develop than planned. The main engines for the orbiter were failing tests. The process of bonding reentry heat protection tiles to the shuttle surface was turning out to be far more difficult than had been anticipated. The computer models used to simulate the shuttle flight were not yet trustworthy.[123] NASA adjusted the dates for which it would continue to make expendable rocket launches available.

The shift in dates was slight compensation for satellite owners. Their problem was uncertainty—uncertainty as to whether they should budget for the cost of a shuttle launch or for the more expensive cost of an expendable booster launch, and uncertainty over when their "bird" would actually get off the ground. NASA was also beginning to raise its estimates for the cost of a shuttle flight (estimates that satellite makers had always viewed with skepticism) and to lower its projection for the number of flights per year.[124] To add to the dilemma, NASA had stopped funding upgrades in the Delta to match the continual rise in satellite weights. By the late 1970s, some of the satellites coming on line were too heavy for the Delta launchers then available.

Industry leaders and Congressional representatives joined forces in 1979 to strong-arm a reluctant NASA into participating in the ordering of another Delta upgrade.[125] This solved some of the problems for satellite owners like Indonesia or Telesat Canada. It did not bail out Hughes, however, whose shuttle-sized satellite—by 1980 being called Leasat—with a diameter of fourteen feet could only be launched in the shuttle. In 1980, it was a distinctly mixed group of blessings that the shuttle brought to the communications satellite industry.

Space and the Marketplace

N JANUARY 1981, RONALD W. REAGAN MOVED INTO THE
White House. In April the space shuttle *Columbia* made its maiden,
two-day, orbital flight, demonstrating that after two years of delay, the
shuttle was finally coming on line. In June, the European Ariane rocket was
launched for the third time from a base in French Guiana, South America, and successfully placed in orbit a European weather satellite and an
Indian experimental communications satellite.[1] It was clear that in the Ariane the shuttle was going to have bona fide competition for foreign and
commercial payloads.

The problem was exacerbated by Reagan's victory because the new
president brought with him the conviction that economic activity, wherever possible, should be the province of the private sector.[2] The shifting
of U.S. government launches to the shuttle would mean that the firms that
made expendable launch vehicles would no longer have a government
market for their boosters. The Reagan administration immediately raised
the question of whether these companies should be permitted to offer expendable launch vehicles to nongovernment buyers on a commercial basis. This, however, would place NASA in the position of having to compete with portions of the domestic aerospace industry as well as with
Ariane. As part of the administration, NASA was under obligation to help
implement Reagan's support for private initiative. But because making the

shuttle cost-effective depended on securing nongovernment payloads for it, NASA would be in direct competition with the private launch operations it would ostensibly be furthering.[3]

For their part, the expendable launch vehicle producers had to ask themselves whether it made more business sense to close down their production lines or convert them to commercial use at the very time when the largest market, that of U.S. government launches, was slated to be reserved for the shuttle. These companies—McDonnell Douglas, Martin Marietta, and General Dynamics—were large, diversified firms with NASA contracts. They required a strategy that would not only maximize their launch vehicle profits, but would also protect their positions as NASA contractors. They had, as well, to put into the balance those new business opportunities that were opening up. The Reagan government was ramping up the budget for military space, creating prospects for space business in that area. And the aerospace community was watching Reagan's NASA administrator, James M. Beggs, and his Deputy, Hans Mark, as they tried to win approval for a space station program that would entail multibillion dollar contracts.

Meanwhile, the new administration's pro-business philosophy helped inspire the entrance into the space industry of several dozen new entrepreneurial firms. Most of these companies proposed products or services that would complement the shuttle, but a few of them intended to market launch vehicles that would compete with it.

Entrepreneurial Firms

Entrepreneurial firms were drawn into space business in the late 1970s and early 1980s for several reasons. The very reconceptualization of the shuttle served as encouragement. When the shuttle was primarily part of a grand scheme to explore Mars, it seemed natural that space logistics remain with the government. But when it metamorphosed into an inexpensive way of putting satellites into orbit, it fell into the general category of modes of transportation. Other modes, like the railroads and airlines, were in the hands of private companies. Why not space transportation?

Commercial communications satellites had proved successful. High-technology enterprises had become increasingly popular with venture capitalists during the 1970s. Reaganites were extolling free enterprise in general and small business in particular. An optimism by no means limited to Republicans had been engendered by the shuttle and Ariane. There was a sense that a new era was dawning in space, one in which commercialization would play an important part.[4]

Even before the shuttle flew, a number of men had set up firms to design and manufacture small rockets. Because the first flight of the shuttle had been put off, frustrating satellite owners, the new space entrepreneurs reasoned that the shuttle was creating a demand it was not going to be able to meet. There were, moreover, markets that, from the outset, looked as if they would not be well-served by the shuttle. Small satellites was one. These spacecraft would have to be bundled together, if launched by the shuttle, so as to insure a full cargo bay. Small satellite owners would thus be denied the possibility of choosing a launch time and an orbital inclination that would be most suitable for their crafts.[5]

After the *Columbia* made its maiden flight in April 1981, and even more after it made its second flight in November 1981, a second group of entrepreneurs saw openings. The two flights together demonstrated what could never have been known beforehand with complete certainty, that the shuttle worked. It flew the way that it was supposed to fly and was, indeed, reusable. This group of entrepreneurs saw the chance to build businesses that could offer services or auxiliary hardware for the shuttle.

Some of the new entrants into the space industry had precise business plans. Others were prepared to go after every market that presented itself. Some were former NASA engineers and managers. Others were outsiders. But the outsiders almost always brought in people who had worked for NASA at an early stage in the organization of their companies. They wanted the cachet the NASA name could bring, the technical expertise of trained NASA engineers, and the access to the agency that hiring someone from its ranks might provide. Years later, a retiring NASA executive would retell the joke: "[T]he definition of an entrepreneur here in Maryland was somebody who used to work for the government and now sells to it."[6]

A few case histories will give concreteness to these generalizations. David Hannah, Jr., a Houston real estate developer, entered the space business in 1980 after he met Gary C. Hudson. Hudson was a former employee of the Jet Propulsion Laboratory who was developing an inexpensive kerosene-fueled rocket. Hudson's vision of building simple rockets, using off-the-shelf components so as to lower the cost of space transportation, convinced Hannah to organize Space Services, Inc. (SSI) with money from friends and associates in real estate, oil, and other non-aerospace sectors. SSI's first venture was a cooperative undertaking with Hudson's company, GCH, Inc., to finance and build Hudson's Percheron, a rocket that used kerosene and liquid oxygen as its propellants. Hannah arranged to have the Percheron launched from the ranch of one of SSI's investors on Matagorda, an island off the coast of Texas, in August 1981. But the Percheron blew up on the launch pad, and with it went the Hannah-Hudson partnership.

SSI now turned to another California firm to build for it the Conestoga I solid-motor rocket, using surplus parts of a Minuteman rocket purchased through NASA from the military. Conestoga I had better luck. It flew from Matagorda in September 1982, executing a 320-mile parabola over the Gulf of Mexico, and successfully releasing its test payload, a module filled with forty gallons of water. Conestoga I received wide media attention, for it was the first successful rocket to be sent up by a commercial company.

Space Services quickly acquired NASA personnel. Donald K. (Deke) Slayton, a former astronaut, was brought in to direct the Conestoga I launch. He soon moved up to become company president. SSI also made use of the considerable number of NASA engineers who were reaching retirement age in the early 1980s. Hiring them as consultants on an hourly or daily basis gave the company inexpensive access to a deep pool of expertise.

Space Services Inc.'s targets for markets were fluid even during the company's formative years. Hannah's fundamental premise was that if access to space were made sufficiently cheap, then new uses for launch vehicles would develop, just as new and unexpected uses had developed for semiconductor chips as their costs had dropped. At first, SSI spoke of lofting small payloads—three hundred to five hundred pounds—into low earth orbits. This was a niche market that would complement, rather than compete with, the shuttle. But in 1982, SSI talked of eventually launching large commercial satellites into geosynchronous orbit, and by 1983, it projected the building of remote sensing satellites as well as launchers. This would put SSI in an ambiguous position vis-à-vis the National Oceanic and Atmospheric Administration's (NOAA) Landsat. In later years, SSI was to talk of taking over Landsat, of running NASA's smallest rocket, the Scout, and even of sending cremated remains into orbit, so that the loved ones of its customers might find their physical, as well as spiritual, resting place in heaven.[7]

Arc Technologies (later Starstruck, Inc.), is another firm established in order to build commercial rockets. James C. Bennett, one of its founders, argued passionately against leaving space transportation in government hands. The outcome could only be inefficient and costly systems. The shuttle, according to Bennett, was a one-size-fits-all spaceship and this flew in the face of experience already gained in land, sea, and air transportation. In these earlier technologies, customer needs were met by a variety of vehicles, each designed to sell to one or two markets.[8]

The established aerospace firms did not earn much more respect than the government did from Arc Technologies-Starstruck executives. They were "lumbering aerospace giants" who were too spoiled by their cozy contractor relation with the government to function well in a commercial arena. What was needed, rather, was the light-on-your-feet entrepreneurship characteristic of the computer firms of Silicon Valley. The company's founders recruited Michael Scott, a former president of Apple Computer, to be company president. They raised most of the capital from Scott himself, from Stephen Wozniak, one of Apple's founders, and from other computer executives.

In addition to a different style of management, Starstruck based itself upon a different technology. Its Dolphin, a reusable fifty-foot rocket, was to be launched from water. This would allow it to be towed to favorable launch latitudes, although it carried the penalty that equipment failures were made harder to fix. It was to be powered by a "hybrid" rocket motor with a solid fuel and liquid oxidizer. This combination made manufacturing safer and therefore cheaper than the manufacture of solid rockets, where the fuel and oxidizer are together and can explode during casting. It eliminated the inconvenience of liquid hydrogen, which must be kept more than 100 Fahrenheit degrees colder than liquid oxygen. The absence of liquid hydrogen, and the fact that the burning could be cut off by shutting down the flow of oxygen, made launching safer too.[9]

While these firms worked to field rockets, others sought to market services that would support the shuttle. Astrotech Space Operations was started by a man with a thorough knowledge of NASA's ways: Robert J. Goss had been head of the payload integration branch in the Delta rocket program at Goddard Space Flight Center. He retired from Goddard in 1980, and started his own consulting firm, Astrotech International Corporation.

NASA had changed its billing practices for commercial payloads in moving from expendable launchers to the shuttle. Its prior practice had been to include the processing of payloads with other launch services. Now it broke off payload processing, charging a separate one million dollar fee for it. The new arrangement raised some discontent among commercial customers. Goss saw an opening here since he believed a commercially operated processing facility could do the job at a much lower price. In 1981, he began the slow process of convincing NASA to allow Astrotech International to set up a private facility outside Kennedy Space Center to augment NASA's own processing facilities. At the same time, Goss worked to convince satellite makers that Astrotech could do the job competently and reliably, and for less money than NASA.

Astrotech achieved a close working relation with NASA and by 1983 it had procured a site and designed a processing facility. It had not, however, found investment capitalists who were sufficiently venturesome to advance funding for construction. The man who came to the rescue was Willard Rockwell, who had retired from Rockwell International in 1979. Together with some associates, he had just bought the Cyprus Corporation, an investment firm, with the intention of transforming it into a high-technology enterprise. Cyprus had begun buying small, high-technology companies, and it bought Astrotech International. Goss got his funding through this purchase, but Rockwell got the name: Cyprus was renamed Astrotech International and Goss' company became Astrotech Space Operations, its subsidiary. In April 1984, Astrotech Space Operations opened a $7.5-million installation at Titusville, Florida, three miles from the Kennedy Center. By early 1985, it had signed contracts with all the shuttle's 1985 commercial customers.[10]

Orbital Sciences Corporation was the brainchild of three men who had been classmates at the Harvard Business School. David W. Thompson was an aeronautical engineer who had worked briefly for the NASA Marshall Space Flight Center in the late 1970s. There he had become convinced that a private company, wholly dedicated to space, could be more effective in developing the new technologies "than either a government agency or a large aerospace company that viewed space as a sort of sideline to its primary business of building defense systems or commercial aircraft." That idea sent him to the Business School at Harvard, where he met Bruce W. Ferguson, who was studying law and business, and Scott Webster, who, like Thompson, had a background in engineering.[11]

Harvard Business School had the practice of inviting industry to propose problems for a course called Creative Marketing Strategy. The students organized into teams and each team received a grant from the company whose problem it attacked. Thompson, Ferguson, and Webster joined a team working to identify commercial opportunities in space. It was funded by Marshall's Program Development Group, as part of its project of developing support for materials processing in space. In 1981, the three men, now graduated, met in Houston to receive an award for their Harvard Business School study. There, they decided to found Orbital Sciences Corporation (OSC).

OSC first proposed to NASA that it take over the responsibility for a shuttle-compatible Centaur upper stage rocket for lifting the heaviest class of satellites to geosynchronous orbit. When NASA demurred, the firm changed to a different product, a family of upper stage rockets for medium-weight satellites. OSC emphasized that these rockets would not compete with NASA. On the contrary, their plan would provide an infusion of private funds that would make it possible for NASA to pursue its shuttle program more effectively. Orbital Sciences contracted with Martin Marietta to develop and manufacture the upper stages, and with NASA's Huntsville Center to provide technical supervision, and secured an initial $1.8 million in venture capital. Then it went out to raise the fifty million dollars it needed to finance production. "Martin Marietta," Bruce Ferguson later said, "wanted a contract in cash without the risk: we provided that. NASA wanted an upper stage without spending money out of its budget: we provided that." As for the three founders, "Dave, Scott, and I believe strongly in the importance of trying to commercialize the space industry."[12]

Each of the firms I have discussed was paralleled by others. Robert C. Truax, a retired naval officer with many years of rocket work in the military, raised one million dollars in the late 1970s to build small rockets, including a "Volksrocket" for flights by ordinary citizens. GTI Corporation of San Diego signed an agreement with NASA in 1982 to go into business as an intermediary between NASA and firms unused to dealing with government agencies. John M. Cassanto founded Instrumentation Technology Associates to build space-qualified instruments for companies wishing to do research in space. Commercial Cargo Spacelines proposed to charter whole shuttle cargo bays and rent space on them for satellite launches and experiments.[13] David W. Thompson, president of Orbital Sciences Corporation, estimated in 1984 that the number of companies in commercial space enterprises had grown from three in 1980, with total investment of $10.5 million, to twenty-five in 1983, with total investment of about $175 million.[14]

A MISSIONARY zeal for bringing capitalism to space infused many of the leaders of these companies. Often it took the form of championing private, as against government, enterprise; sometimes it exalted small, entrepreneurial firms as against large aerospace companies. Free enterprise, claimed these men and women, could create technology that was better, cheaper, and more suited to market demands. A nationalistic element entered. They argued that free enterprise was a part of U.S. tradition, so that to promote it in space was to defend the American way.[15] They held as an article of faith that small firms were more innovative than large ones.[16]

The companies freely acknowledged that much of the technology they were employing, or the expertise they availed themselves of, had been created with taxpayer money. The surplus rocket engines SSI bought through NASA for its Conestoga I had been part of an Air Force Minuteman missile. The shuttle that Orbital Sciences and other companies sought to complement had been developed by NASA. The market research that inspired Orbital Sciences had been done on a NASA grant. The NASA retirees brought in as consultants or as company officers had acquired their expertise on government payrolls. To counter arguments against allowing public expenditure to benefit private companies these executives praised the precedents of the railroads and the airlines and interpreted them as showing that the U.S. has always had the wisdom to spend public money to establish private sector enterprise.[17]

The new right ideology animated these entrepreneurs and provided them with concepts they could rest against. George Gilder was just publishing *The Spirit of Enterprise,* an eloquent paean to the small entrepreneur—rebel and creator, altruist and optimist, immersed in the Judeo-Christian faith and culture—the antithesis of the naysaying, desiccated and godless intellectual.[18] On a more down-to-earth level, Reagan's advisors were sanguine about space commercialization.[19]

The spokesmen for space commercialization maintained that the private sector would make better use of resources in exploiting space, because private enterprise is governed by the discipline of the marketplace. Government, in contrast, feeds into its economic decisions noneconomic considerations of domestic and international policy. Yet these 1980s entrepreneurs were patently moved, at least in part, by ideologies and enthusiasms that had nothing to do with the bottom line.

The Office of Commercial Programs

As the number of commercial space firms multiplied, the need increased for NASA to reorganize itself to deal with them in an efficient and coherent manner. An impetus toward such a reorganization was provided by a new national space policy, released on Independence Day 1982, as President Reagan spoke in celebration of the landing of the fourth *Columbia* shuttle flight. The policy affirmed that "The United States encourages domestic commercial exploitation of space capabilities, technology, and systems for national economic benefit . . . consistent with national security concerns, treaties, and international agreements. . . . The United States Government will provide a climate conducive to expanded private sector investment and involvement in space activities."[20]

On the heels of this policy release, Administrator Beggs' requested the National Academy of Public Administration to study how NASA might comply with the president's policy. The panel the academy convened, headed by former Secretary of Commerce Philip M. Klutznick, concluded that in the current disposition, commercialization activities were badly fragmented among the various agency offices. Negotiations with companies that wanted to get into the space business were outrageously protracted. It had taken Microgravity Research Associates from 1979 to 1983 to negotiate a joint endeavor agreement with the agency, and that company's difficulties were not atypical.

The panel opined, moreover, that NASA had an R&D mind set that made considerations central to commercialization, like cost, market demand, and financial risk, foreign to it. All this argued for the establishment of a separate NASA office under managers with marketplace experience. The panel also advocated that NASA establish R&D facilities to serve commercial space, much as the old NACA had used its wind tunnels to do generic research beneficial to commercial airplane manufacturers, and to troubleshoot specific aircraft. Advisory committees with industry representatives would suggest research problems, again much like NACA procedures.[21]

Beggs set up a Commercialization Task Force in June 1983 to study the study. Meanwhile, Congress had gone on record as supporting a commercialization office within NASA. NASA itself had begun to be more interested in private space business, as it began to push for presidential approval of a space station; use of the station for commercial projects was one of the justifications NASA was offering. In September 1984, the Office of Commercial Programs (OCP) was formally inaugurated. The man chosen to head it, Isaac T. Gillam, IV, had worked for much of the 1960s and 1970s in that part of the agency that could most nearly be called a business, selling launch services on Delta rockets to foreign governments and private firms. The budget for the office for its first year, FY 1985, was $17.1 million, about 0.2 percent of NASA's 7.573 billion dollar budget.[22]

The Office of Commercial Programs had a full menu of tasks. It was to simplify and accelerate the process of negotiating agreements between NASA and private firms. It was to serve within NASA as an advocate for space commercialization and between various NASA offices and industry as a broker to reconcile conflicting interests. It would develop criteria for selecting commercial projects that would be at once likely to succeed and not likely to conflict with government needs or obligations. It was to administer the new Centers for the Commercial Development of Space that Beggs had begun speaking about early in 1984. These would be university-based institutes, funded by NASA but with active participation, either through financing or other types of contributions, by private industry. They would work to assemble the kind of data that commercial companies could use to decide on the profitability of space ventures. The office would also sponsor intramural, NASA research programs to generate additional data.[23]

The office faced the immediate problem that many of the functions it had been awarded had been pulled from other programs, whose staff was not always keen to relinquish them. It ran into territorial conflicts over research into materials processing in space with the Office of Space Science and Applications, which had housed such research since the 1970s. It found itself in quarrels with the Space Shuttle Office over Beggs' assignment to it of authority for marketing shuttle payload bay space.[24] It had the duty of encouraging nonaerospace companies to interest themselves in using a NASA-built space station. But so did the Space Station Task Force, so once again the office was involved in divvying up tasks.

The office faced the more profound problem that the commercial projects it was supposed to broker required cooperation and even funding from NASA's other program offices. To fit a piece of commercial apparatus into the shuttle cargo bay, for example, might require Johnson Space Center engineers to adjust other aspects of the shuttle's mission. To award industry fly-now pay-later contracts to send up commercial equipment could disrupt the cash flow of the headquarters shuttle office.

There were also genuine differences in the objectives of the commercial companies and the NASA offices that the OCP was charged with bringing together. NASA engineers, whose job it was to guarantee the safety of astronauts, needed to know every feature of the equipment they were asked to integrate into systems that would take humans into space. But industry managers wanted to keep new ideas confidential, so as to preserve their proprietary secrets. Further, NASA engineers were comfortable taking one, two, or three years to integrate a new device into the shuttle. Business, especially small business, did not have the money it took to keep a research project alive over such long time intervals.

At times, a simple tug-of-war over who would get to do a program would ensue. Wrote one of NASA's Advisory Council Task Forces, "Typically a proposal is made and encouraged by upper NASA management. When the lower level managers and specialists review it, they tend to see opportunities to fulfill the need as an in-house program and therefore kill or maim the proposal."[25]

The Office of Commercial Programs worked hard during its first year to develop a range of contractual instruments with which it could support companies during the different stages—research, development, early commercialization, maturity—that it took to bring a new product to market. It took over the often difficult negotiations then in progress between NASA and companies seeking to privatize NASA systems or to explore space applications. Its Centers for the Commercial Development of Space did important research on space applications. But OCP was a small office with a staff of thirty to forty, and a low budget. It had to contend with indifference in the business community and a firmly rooted R&D culture within its own agency. It faced formidable problems in getting "all of the [NASA] heads rocking up and down [so as to] get a yes out of the agency."[26] In the end it was, at best, a new procommerce voice within the cacophony with which NASA spoke to the private sector.

Space Transportation and the Established Companies

While the small companies got started, the "lumbering aerospace giants" were also examining the new landscape. They looked at the same niches in space transportation the small firms were trying to fill: marketing services for the shuttle, processing payloads for the shuttle, upper stage rockets to transfer payloads to higher orbits, and commercial launch vehicles. But the large companies operated under a more complex set of conditions. Unlike the start-up firms, most had some sort of shuttle contract. They had, as well, other prospects that the small companies could not aspire to. NASA was then in the process of seeking administration and congressional approval for a major new project: the space station. It would be the first new start in the field of manned space flight since the shuttle contracts a decade earlier. Another burgeoning market was the Department of Defense, whose procurements for space had been growing since the mid-1970s, and, beginning in fiscal year 1982, overshadowed NASA's. Moreover, President Reagan was in the process of adding another military space venture, the Strategic Defense Initiative, to the federal budget. We need to examine these alternative markets more closely to understand the established companies' strategies on launches and shuttle-related hardware and services.

The space station had been on the minds of Ronald Reagan's appointees for NASA Administrator and Deputy Administrator, James M. Beggs and Hans Mark, since they had taken office in July 1981. Both had come to their jobs with a deep commitment to elevating it from a perennial dream to a core NASA program. As one after another of the four shuttle test flights was successfully completed in 1981 and 1982, they took steps to bring the station into being. They created a small Space Station Task Force at headquarters, under John D. Hodge, a former NASA and Transportation Department engineer. And they began the process of getting presidential approval for the project. This latter was by no means a sure thing. David Stockman, Director of the Office of Management and Budget, opposed the station. So did George A. Keyworth, Reagan's Science Advisor, Caspar Weinberger, the Secretary of Defense, and many high officials in the Air Force.[27]

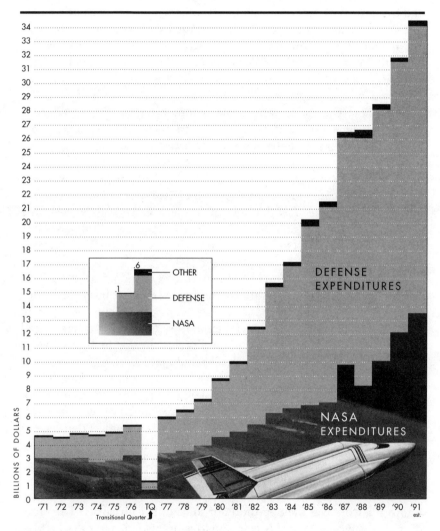

NASA versus Department of Defense space expenditures. The figures are in current-year dollars. They may not add because of rounding. Department of Defense expenditures on missiles are not included.

Source: Aeronautics and Space Report of the President: 1989–1990 Activities, Washington, D.C.: NASA, 1991, 162.

NASA strategy was first to elaborate the scientific, commercial, and military missions to which a station might be put, deferring discussion of its design. Eight aerospace companies received small contracts in August 1982 to aid with the definition of the station's missions. Boeing, Lockheed, TRW, Grumman, Rockwell, McDonnell Douglas, Martin Marietta, and General Dynamics got just under $800,000 each for the studies. One of the few details that were firmly stipulated was that the components out of which the station and its satellite structures were to be built would be carried into space by the shuttle, rather than by expendable launch vehicles.[28]

Beggs wished to involve as many centers as possible in the station. This would give it a wide constituency within NASA, and it would enlist the congressional representatives of the districts in which the centers were located. Marshall and Johnson were both actively seeking as much of the program as they could get and were each contending for the role of lead center. Goddard was not interested. It was the center most closely linked to the scientific community, however, and Hodge wanted Goddard in the mix to give scientists confidence in the project. Lewis was NASA's core center for space propulsion, and Lewis' director, Andrew J. Stofan, wanted the station's power system, so as to add to the center a new and related area of expertise.[29]

To the aerospace industry, the station had many attractions. They saw manned space as a natural extension of the aircraft business. Transporting people through space was the next step beyond transporting them through the air. The habitation and laboratory modules of a station had structural affinities with the bodies of airliners. It was the kind of work that made excellent copy for advertising and that helped attract the most capable members of the new crops of graduating engineers.[30] Finally, and not least, it would mean multibillion-dollar contracts over a decade or more. The companies therefore lobbied Congress and the president on behalf of the station. Meanwhile, they jockeyed to aggrandize the share of those NASA Centers with which they had the closest ties, and from which they were most likely to receive station contracts.[31]

At the end of 1983, Reagan decided to approve the station. His own enthusiasm for space and for the aerospace industry of his home state of California won over the opposition of some of his advisors.[32] The president's announcement of this new initiative, in his January 1984 State of the Union address, was industry's signal that contracts would be coming its way, first for the Phase B stage of defining the station's architecture, and later on for hardware. By this time, NASA had begun to adjust its internal rivalries by dividing the station into four "work-packages," one each to be managed by Johnson, Marshall, Lewis, and Goddard, and by choosing Johnson as the lead center for Phase B.[33]

During the years that the space station took its tortuous route to presidential acceptance, the market for military space products grew lushly. Satellites that could aid ground warfare became an increasingly important part of the arsenal. Use of such satellites had begun with the reconnaissance spacecraft the Air Force had first procured in the 1950s, and had gone on to include communications satellites, meteorological satellites to ascertain battlefield weather conditions, navigation satellites to help planes and ships fix their positions, and sensor satellites to give early warning of enemy missiles.[34]

Military expenditures for satellites actually fell in the early 1970s as the satellites became more capable and longer-lived.[35] But procurements had picked up since 1977. A renewal of Soviet tests of antisatellite weapons helped to spark concerns about the vulnerability of the military's now-indispensable satellites. Spending grew for measures to increase satellite survivability, test antisatellite weapons, and perform preliminary research into weapons using laser or particle beams.[36]

President Reagan, who raised the defense budget across the board when he came to office, also poured more money into space systems.[37] Defense Secretary Weinberger's Strategic Modernization policy gave added emphasis to the whole field of space weaponry. Meanwhile, the Air Force and Navy raised the stature of space within their organizations by creating Space Commands.[38] The contracts that came out of these programs commanded attention across the aerospace industry. In March 1983, President Reagan announced his Strategic Defense, or "Star Wars," initiative. It was a racheting up of the work being done on particle beams and lasers with the specific objective of using them as antimissile weapons. The cost of first generation Star Wars systems would run $75 to $150 billion. Actual procurement was ten to fifteen years away, but immediate spending, for research and development, would be at high levels, some seventeen to twenty-six billion dollars over five years. Throughout the industry, firms began to put their resources into company financed research projects and facilities, so as to make themselves candidates for Strategic Defense Initiative contracts.[39]

All this time, a steady drumbeat of pronouncements from the administration and Congress in favor of the commercialization of space transportation could be heard. The president followed his July 4, 1982, space policy, which affirmed support for private sector investment in space, with a specific endorsement of a commercial expendable launch vehicle industry in his January 1983 State of the Union address. In May 1983 he issued

National Security Decision Directive 94 stating that "[t]he U.S. government fully endorses and will facilitate the commercialization of U.S. expendable launch vehicles." The directive promised that the government would encourage use of its launch ranges and would sell to the private industry tooling, equipment, facilities, and services at reasonable prices.[40]

In February 1984, Reagan designated the Department of Transportation as lead agency for launch vehicle commercialization in Executive Order 12465. Transportation Secretary Elizabeth Dole immediately set up an Office of Commercial Space Transportation with the twin tasks of facilitating the licensing of private launches and serving as advocate within the administration for the commercial launch vehicle industry. Meanwhile, the administration promised to adjust the shuttle prices after fiscal year 1988 to reflect the full marginal cost of the flights.[41] This would make it possible for commercial launch firms to compete with the shuttle on a cost basis, as well as on the basis of convenience and reliability of scheduling. The Congress had already worried over the difficulties firms like Space Services were having in securing licenses to launch. From 1982 on, it had considered a number of bills to correct the problem. In October 1984, Congress moved to give the Office of Commercial Space Transportation a more secure legal underpinning by passing the Commercial Space Launch Act, Public Law 98-575.[42]

We may now characterize the space market as it looked to the large aerospace companies in the period from 1981 to early 1984. Congress and the administration sought to encourage the growth of a commercial launch industry. But other opportunities abounded. There was NASA's space station. There were DoD satellites. The announcement of the Strategic Defense Initiative in March 1983, and the setting up of the office for the Strategic Defense Initiative Organization, constituted the prelude to billions of dollars of research and development contracts with the potential of enormous hardware contracts down the line. All these systems—the station, new military satellites, and the space-based parts of the Strategic Defense Initiative—would, of course, need transportation into space. But both NASA and the Department of Defense had committed themselves to the use of the shuttle and were not prepared to place their cargoes on commercial expendable launchers.

It is true that some economists were predicting launches of two hundred to two hundred fifty commercial satellites over the next ten to fifteen years. But all but the smallest of these were cargoes NASA wanted for the shuttle. NASA argued that capturing these payloads would make it possible to fly the shuttle often and flying often would bring the cost of space transportation down. It maintained that this would do more, overall, to encourage space commercialization than would ceding part of its cargo to commercial expendable launch vehicles, for it would keep launch costs low for commercial satellites and would accelerate the birth of a space manufacturing industry.[43] Further, not all the large companies believed the high estimates that launch vehicle entrepreneurs and NASA contractors were projecting for commercial satellite launches.[44]

Given these circumstances, it is not surprising that none of the large aerospace companies emulated Space Services or Starstruck and initiated new launch vehicle programs. Only the three companies that were already making expendable launchers—General Dynamics, McDonnell Douglas, and Martin Marietta—grappled with the issue. For them it was a question of whether to convert their lines for commercial products or close them down. Each of the three producers of the larger expendable launch vehicles adopted a different strategy.

McDonnell Douglas, the manufacturer of the Delta, decided not to market it. "In our commercial projects," said a company executive, "we have been careful not to compete with NASA but rather to be compatible with, and companion to, its efforts."[45] This position was not arrived at without some internal controversy. McDonnell Douglas had been selling NASA Deltas for twenty years. It was a small part of the company's output, but a consistent money maker. Theodore D. Smith, the retiring vice president for Space Transportation Systems at McDonnell Douglas Astronautics, believed NASA could never achieve the launch rate of twenty-four shuttle flights per year at the cost per flight of about seventy million dollars that the agency then projected. In 1982 shuttle costs were running about three hundred million dollars per flight and shuttle turn-around times were running upwards of three months. "I give the Shuttle three more years before they admit they can't compete economically for Delta class payload launches," he wrote in the fall of 1982.[46] On the other hand, McDonnell Douglas Astronautics' new president was John Yardley, the hard-driving administrator who, as head of NASA's Office of Space Flight, had shepherded the shuttle through its maiden flight in spring 1981. Yardley had faith in the shuttle and thought that any commercial expendable launch vehicles that were marketed in competition with the shuttle were doomed to failure.[47]

McDonnell Douglas was willing to entertain the idea of building Deltas for another firm, specifically for the entrepreneurial company Transpace Carriers, Inc. (TCI). TCI had been founded by David W. Grimes, a Delta Project Manager at NASA's Goddard Space Flight Center. The impetus was the announcement in August 1982 that NASA would stop launching the Delta in 1986. Grimes promptly resigned to set up his company.

He argued that the Delta's reliability (as opposed to Ariane's launch failures) and its ability to send up payloads on schedule (as opposed to the shuttle's delays) put it in a strong position to win contracts to launch commercial satellites.

In September 1983, following President Reagan's May 1983 directive supporting a commercial launch industry, NASA issued a request for bids for companies to take on the Delta and Atlas-Centaur programs as private ventures. TCI was the lone company to bid for Delta. It concluded a preliminary agreement with NASA providing a period for the newly formed company to find three commercial customers and secure funding. TCI also reached an informal understanding with McDonnell Douglas and went out to search for capital and customers.[48]

Like McDonnell Douglas, Martin Marietta looked into the possibility of commercializing its Titan through an entrepreneurial firm. In Martin Marietta's case, the firm was Space Transportation Company (SpaceTran) with whom Martin Marietta made preliminary agreements in 1982.[49] When, in early 1983, SpaceTran sold off to Federal Express the expendable launch vehicle part of its business, Martin Marietta entered into a joint venture with Federal Express to build launchers. The two companies submitted a proposal to the International Telecommunications Satellite Organization (Intelsat) to launch the Intelsat VI satellite. The other bidders were Arianespace, which proposed to use an upgraded rocket—the Ariane 4, then being developed—and NASA, which bid the shuttle. It was an awkward situation, because Martin Marietta was NASA's second largest contractor for the shuttle. The company argued it was competing with Ariane, but complementing the shuttle since, it reasoned, Intelsat was bound to want to place some of its new satellites on the shuttle.[50]

The Martin Marietta–Federal Express bid did not prevail over the government subsidized prices of the shuttle and Ariane. In the fall of 1983, the joint venture was dissolved. It looked as though Martin Marietta would close down its Titan launch vehicle production line for good in 1986, when its last booster for the Air Force was due to be completed.[51]

Of the three launch vehicle companies, General Dynamics was the only one that was unequivocally committed to commercializing its Atlas-Centaur booster, and it proposed to do so alone. The company began its assessment of market possibilities in 1981 and by April 1982 announced its intentions to NASA and the Defense Department. By January 1983, it submitted a formal proposal to NASA. When NASA issued its call for companies to take over the agency expendable launcher programs in September 1983, General Dynamics was the only firm to bid for Atlas-Centaur.[52]

To Congress, the company presented a list of government actions that would help it make commercial Atlas-Centaurs a success. General Dynamics wanted firm administration and congressional statements in support of a commercial launch industry. It wanted to be able to rent government launch facilities at a price limited to the additional costs its launches would impose on those facilities. It further wanted a good price for any spare parts or production equipment it would buy back from NASA and the Air Force. And it wanted NASA to take on the job of doing basic research on expendable launchers, so as to provide the launch industry with the same kind of R&D support the old National Advisory Committee for Aeronautics had provided the airplane industry.

The Atlas-Centaur had launched all the Intelsat IV satellites, but had lost some of the Intelsat Vs to the Ariane. General Dynamics' pitch to Congress was similar to what Martin Marietta was saying about the Titan: To help the commercial launch business was to preserve market share for U.S. firms in the head-to-head fight between this nation and foreign competitors.

IN MARCH 1984, the Department of Defense reversed itself. It abandoned its commitment to exclusive use of the shuttle and sought to purchase expendable launch vehicles. The shuttle was proving far more costly and troublesome to bring into routine operation than Defense had anticipated, and the inertial upper stage that was to take DoD systems from the shuttle into geosynchronous orbit turned out to be expensive and cranky. There were additional reasons why DoD feared to rely solely on the shuttle. There might be military crises in which it would be foolhardy to send up Defense satellites on manned vehicles. The shuttle might need to be grounded and satellites had become so integral a part of the American arsenal that an unexpected inability to launch them could jeopardize U.S. military posture. Finally, DoD was uncomfortable letting the launch of sensitive military payloads depend on a government agency (NASA) that was outside the Defense Department.

Earlier the DoD had hoped an emerging commercial launch industry would be the answer. If some catastrophe were to overtake the shuttle, the existence of commercial Titans would give the department a means to send up its most urgent cargoes.[53] By 1984, however, the Pentagon no longer felt it could rely on a commercial launch industry for back-up. Already in 1983, the Department's Aerospace Corporation had warned that the commercial satellite market would not sustain a private expendable vehicle industry, given the competition that Ariane and the shuttle were offering. If the government wanted to keep expendable launcher lines open, Aerospace argued, it would have to purchase some vehicles itself. Some months after that report came Federal Express and Martin Marietta's failure to sell the Titan to Intelsat. This concrete failure to commercialize an expendable launch vehicle may have played some role in Defense's decision.[54]

Beggs fought like a tiger against the Defense Department's action. He argued that its effect would be to subsidize industry to keep boosters that were already obsolete in production. It would be a mistake to use such boosters as a basis for commercial lines. Removing Department of Defense cargoes from the shuttle, moreover, would jeopardize NASA's project of making the shuttle cost-effective, and jeopardizing this project would threaten the nation's ability to retain a major share of the international market for launches.[55]

When, in the end, NASA's arguments failed to prevail, the agency did something extraordinary. It decided to submit a proposal of its own in response to Defense's request for proposals, one based on the shuttle. This was the so-called Solid Rocket Booster Experiment, the SRB-X, and it was one of several ideas grounded on the shuttle that NASA had been contracting out for study. The SRB-X was based on the solid rocket booster that Thiokol was manufacturing for the shuttle. Beggs pointed out that SRB-X would be a more advanced booster than the Atlas-Centaur or Titan. At the same time, its use would make manufacturing the shuttle cheaper because the fixed costs of the Thiokol booster could be spread over more units.

The submission of a bid from one government agency to another was so unusual that the Pentagon was not even sure of the legal procedures it had to follow to handle it. Nor did it much like NASA's submission. It was a mere concept, without any basis except paper studies. NASA's two competitors, Martin Marietta and General Dynamics, in contrast, were submitting proposals for upgrades of two boosters with long records of success. Industry spokesmen, for their part, were aghast and exasperated. "It is a case of government competing with government and industry is in the middle," said one, while another darkly suggested that it was a move to keep industry from working directly on Air Force space projects.[56] The whole acrimonious episode stood as a vivid example of NASA's determination to protect the shuttle. It was a fight the agency lost: in early 1985, the Defense Department contracted with Martin Marietta for ten upgraded Titans, for use starting in the late 1980s.

In Search of NASA's True Role

Even as NASA battled competitors to the shuttle it debated whether it should be running the shuttle at all. Many of NASA's leaders saw the agency as an engineering research organization and its fundamental mission as the development of new technology. In their view, NASA did not have what it took to carry out routine operations. Either NASA personnel would convert the routine into the challenging by indulging in their engineers' impulses to make unnecessary improvements to the shuttle, or routine operations would drain resources from R&D and corrupt NASA's pioneering mentality.[57]

From the late 1970s on, a series of reports had discussed the pros and cons of alternate arrangements for ownership and operation of the shuttle: operation by NASA, by a group within NASA that was walled off from the rest of the agency, by another government agency or department, ownership and operation by a new government corporation, by a private firm that was government-regulated and government-subsidized, or by a private, regulated, but unsubsidized firm.[58]

Turning government operations over to private firms was much in vogue in the early 1980s. In Great Britain, Prime Minister Margaret Thatcher had embarked on a wholesale privatization of government functions.[59] In the United States, conservative scholars were advocating privatization on both a local and federal level.[60] As far as space went, President Reagan had accelerated the process, started under the prior Carter administration, of turning the government's Landsat earth-observing satellite system over to the private sector, in this case to EOSAT, a partnership of Hughes Aircraft and RCA. He was also pressing for privatization of the National Oceanic and Atmospheric Administration's weather satellites.[61] An advisory committee that he appointed and put under the chairmanship of J. P. Grace to examine ways to cut the federal budget advised passing legislation "that would: allow the private sector to purchase and operate the fifth shuttle; [and] create the option for the private sector to purchase and operate future additional shuttles."[62]

One problem, however, was that the shuttle was far from ready for regular operation. It is true that in July 1982, at the end of the four test flights, Reagan had proclaimed it "fully operational, ready to provide economical and routine access to space."[63] But two year later, the launch rates were still in the single digits, far short of the twenty-four flights per year NASA was then projecting as normal. The shuttle flew four missions in 1983 and five in 1984. There were also persistent problems with the orbiter's main engines. In June 1984, for example, the *Discovery*'s maiden voyage had to be aborted when a valve in the engine failed.[64] In March 1985, a NASA committee concluded that only the external tank and solid rocket boosters qualified as operational. They judged the orbiter to be in a state of transition from R&D and projected it would reach its "full capabilities" by the late 1980s. The main engines of the orbiter still lacked sufficient operating margins and longevity. The committee concluded that "the suggestion that the Shuttle should or could be transferred, sold, or leased to private enterprise is simply not a viable alternative for . . . five to ten years."[65]

James Beggs thought that NASA should eventually spin off the shuttle, either to some sort of quasi-public corporation, or to the private sector. "In the long term," Beggs told an interviewer from the National Space Institute, "NASA is not an operating agency . . . agencies with large operational responsibilities inevitably neglect their R&D responsibilities."[66] But in 1984 he projected 1988 as the earliest date at which the shuttle would be ready to be turned over to another organization.[67]

Inside NASA, moreover, there was interest in holding on to the shuttle. First, NASA had hoped to make money selling commercial companies and foreign governments space on the shuttle, money it could plow back into research payloads. Second, NASA was receiving a predictable, and welcome, four billion dollars a year for running the shuttle. Third, the shuttle was a highly visible program. Without it, NASA stood to attract less attention and hence less funding. Finally, in some parts of the organization, there was a gut-level sense of ownership, a view that the shuttle was an inalienable part of NASA's institutional fabric.

Despite these circumstances, and despite considerable opposition by the Air Force and within Congress, two companies did make an attempt to partly own and operate the shuttle in the early 1980s. The first was Space Transportation Company, founded by Klaus P. Heiss, an economist who had coauthored the Mathematica's studies of the shuttle's cost-effectiveness for which NASA had contracted in the early 1970s. Heiss was motivated as much by his convictions about the economics of space transportation and the wish that the United States maximize its competitive position in space as by the desire to turn a profit. In 1978 the Carter administration had decided to limit the shuttle fleet to four orbiters. Heiss thought this was wrong. It meant that Rockwell would close down its production lines in about December 1984, when the fourth orbiter was due to be completed. This would put an end to the kind of gradual improvements and modifications that are facilitated when a system is in continuous production. It would make it difficult to maintain an adequate inventory of spare parts for the orbiter at acceptable prices. Heiss founded Space

Transportation Company (SpaceTran) in 1979 to raise private money for a fifth orbiter.[68]

The question of whether the government should fund additional orbiters was reopened when the Reagan administration came to power. Nevertheless, in February 1982, Heiss approached NASA with a proposal to purchase, and own, a fifth orbiter, that NASA would operate. SpaceTran would take over all the marketing of shuttle cargo-bay capacity to commercial firms and foreign governments and its return on investment would come from the margin between what it took in from these customers and what it paid NASA for launch facilities and operating services.[69]

Private marketing of the shuttle was attracting considerable interest at this point, both inside and outside NASA. Administrator James M. Beggs had ordered "a hard look" at how to transfer marketing out of NASA in late 1982. The NASA Advisory Council recommended it be considered in 1983.[70] NASA, primarily a research organization, had a structure that did not make selling easy. The shuttle's rival, Ariane, in contrast, was run by a company, Arianespace, organized along commercial principles.[71]

Although having SpaceTran market the shuttle might have been attractive to NASA, another feature Heiss was proposing, using expendable launchers as backups, was not. Heiss later reported that when he went to private investment houses to line up funding they insisted on back-up launchers. "We could not assure investors that the Orbiter[s] would always work and be available for non-U.S. government missions . . . And 'insurance' is inadequate compensation for business opportunities left stranded on the ground." Heiss chose Martin Marietta's Titan as the most sensible backup rocket to invest in. It could be upgraded to handle payloads almost as wide as the shuttle took and it could offer a capability in payload weight that Ariane did not yet have. In the fall of 1982, SpaceTran entered into an understanding with Martin Marietta, and its Titan subcontractors, to buy four Titan 34Ds, were NASA to accept the SpaceTran proposal. Here Heiss was on dangerous ground. The Titan was precisely the kind of expendable launch vehicle NASA saw as competition for the shuttle. But Heiss argued that the essential competition was not between the shuttle and U.S. expendables, but between U.S. vehicles and Ariane. The proper goal of U.S. policy should be to secure for American interests the maximum possible share of the world-wide launch market. Then both the shuttle and the private companies would be able to enlarge the number of their customers.[72]

NASA treated SpaceTran's proposal gingerly from the start. The usual questions were asked. Was the shuttle ready to be marketed in 1982 and

1983? At this point its capabilities had not yet been tested and it was not yet clear how much of its capacity would be claimed by the Department of Defense. What if SpaceTran should go out of business? Small companies had a habit of failing. Would new legislation have to be pushed through Congress to allow the agency to make a deal? NASA officials also worried that the arrangement might be one-sided. Were such a deal to be consummated, SpaceTran would take in roughly three hundred million dollars a year in the launch of commercial cargo. What would the government get?[73]

The negotiations went on through 1982 and 1983 but NASA and SpaceTran did not reach an agreement. Heiss later claimed that it was his plan to use Titans that scuttled his project. "NASA made presentations to us that a backup to the shuttle was not in NASA's interest." According to Heiss, NASA told SpaceTran that a commercial Titan might tempt the Air Force to continue using Titans for some of its missions. It might diminish the number of commercial cargoes going to the shuttle, and in the event some mishap were to overtake the shuttle it might endanger the whole program. A NASA official looking back at the failure of the negotiation saw price as the sticking point. When Heiss first put together SpaceTran, it looked as though a just price for an orbiter would be one billion, in 1982 dollars. In 1983 NASA was estimating the price as $2.3 billion. That appeared to be a sum that SpaceTran could not raise.[74]

The other firm that proposed private ownership was Astrotech, whose head Willard F. Rockwell, Jr., was a man given to enthusiasms and large schemes. He suggested to NASA in 1984 that his nascent firm buy two of the orbiters immediately. One would be an orbiter from the existing NASA fleet, the other would be the much discussed fifth orbiter, which Astrotech would order from Rockwell International. Eventually, this would lead to Astrotech International's ownership of the entire system. "We're the only company ready, willing, and able to buy the Shuttle," Rockwell told a *Fortune* journalist.[75]

Astrotech International pointed out that its scheme would keep open the orbiter production line and would provide the fifth orbiter that the administration and Congress were still unwilling to fund. As had Space-Tran, Astrotech proposed to make its money by marketing the shuttle's bay to commercial firms and foreign governments. Both companies projected large demands from such customers. Rockwell's vision for Astrotech—acquiring the entire shuttle system—was grander than Heiss'. Nevertheless, the two men had somewhat similar motives. Heiss had a deep sense of connection with the shuttle, by virtue of having led Mathematica's study of the shuttle's cost structure in the early 1970s. He wanted the shuttle to succeed and, more generally, he wanted the United States to retain a strong share of the international space transportation market. An economist, and a policy wonk, Heiss had definite ideas on how SpaceTran might contribute to these goals.

Willard Rockwell had led the firm that had built the shuttle orbiter for the government. But he did not believe that space should remain the exclusive province of government. Rockwell wanted space commercialization and he was bent on building his little Astrotech International into a high-technology giant at the forefront of a commercial space industry. Contacting politicians at the highest levels, he vetted his proposals with the White House and NASA Administrator Beggs, and left it to the men under him to get them through NASA's middle managers. Those executives spent the next years in discussions with NASA, but failed to reach an agreement. Reflecting on the episode, some of them considered that it had been quixotic to have attempted to pry away a program so fundamental to NASA's self-identity. NASA managers, for their part, point out that Astrotech's proposals simply did not respond to the agency's need to forge agreements that were in conformity with overall government policy. By late 1985, moreover, Astrotech was in financial trouble with high debt and plummeting stock prices, and its ability to raise money to purchase two shuttles was in question.[76]

Policy makers considering the privatization of the shuttle in the early 1980s were interested in the analogies it had with the establishment of the Communications Satellite Corporation two decades earlier.[77] Certainly, there were some similarities. Both cases dealt with technologies in an early stage of development. Both raised questions about whether government or the private sector should own space infrastructure. Both asked whether the fruits of government investment should be turned over to private firms.

There were, however, at least two glaring differences. First, NASA's stake in the shuttle was vastly greater. In the 1960s, it was a matter of maintaining a role in communications satellite R&D. In the 1980s, in contrast, NASA had evolved into an agency organized around core programs in manned space. The desire to maintain such a core had been one reason NASA had wanted the shuttle in the first place, and it made it far trickier to take it away.

The second difference was on industry's side. In the early 1960s, the largest firms in the U.S. communications, electronics, and aerospace industries were all exploring ways to enter the field of space communications. In the early 1980s, industry was scarcely clamoring to acquire the shuttle. The two exceptions were firms so out of the ordinary that their interest serves rather to underline the prevailing lack of interest.

It is true that a few large firms had flirted with shuttle ownership. Boeing had been the most serious. In 1978, its long-range planning group had studied the idea of an industry-wide shuttle consortium, somewhat along the lines of Arianespace. It had rejected the concept on the grounds that the risks were too high and the chances for profits too uncertain. The basic problem was the shuttle itself. Could it meet the launch rates that NASA was promising? Were the government projections of the cost of flights to be trusted? How reliable were the rates that NASA was projecting for commercial and foreign traffic? And how often might nongovernment cargoes be bumped to make room for military payloads?[78] Until these uncertainties were removed, shuttle privatization did not make business sense.

The privatization proposals of SpaceTran and Astrotech were predicated on the use of the shuttle to launch commercial and foreign satellites. In 1986, they would be overtaken by a presidential directive prohibiting the shuttle from carrying such cargoes save in exceptional circumstances.

PARALLELING THE debate over whether to privatize the shuttle was another controversy. What kinds of research should NASA carry out for the commercial space industry?[79] This was a different facet of the fight over NASA's true mission and NASA's work in communication satellite technology became a major battlefield for it. Advocates for one or another view of NASA's proper role in R&D contended for or against the communication satellite program according to whether or not it matched their vision, while organizations with an institutional or financial interest in communications satellites joined the debate by wrapping their self-interest in general pronouncements about the appropriate role for government R&D agencies.

NASA had phased out its work on communications satellite technology when agency budgets were pinched in 1973, with the justification that the commercial satellite industry was now mature enough to carry out its own research. But at Lewis Research Center (which had participated in the Communications Technology Satellite, a successful joint project between NASA and Canada) and in the Communications Division at the headquarters Office of Space Science and Applications there were strong advocates for re-entering the field. Goddard, in Greenbelt, Maryland, was the traditional NASA center for communications satellite work. It had managed the Syncoms and the follow-up series of Advanced Technology Satellites. Lewis argued, however, that its own long history of collaboration with the airplane engine industry under the NACA and NASA better fitted it for a program whose goal was to prove out radical new technologies for commercial use. Goddard, Lewis held, was more attuned to work directed to meeting NASA's internal needs. Goddard had even alienated industry, in Lewis's view, by striving to use advanced technology satellites for purposes like educational broadcasting, instead of having these services supplied through the private sector. In 1979, Lewis won away from Goddard the role of lead center for communication satellites.[80]

Lewis now assumed responsibility for managing a project then taking shape within NASA, an experimental satellite operating at the frequency of 30 gigahertz for the signal from ground station to satellite, and 20 gigahertz for the signal from satellite to the ground (a "30/20 gigahertz" or Ka-band satellite.) This contrasted with the much lower frequencies that operating satellites were then using. Most used 6 gigahertz for the up-link and 4 gigahertz for the down-link (C-band). A few were then venturing into the new territory of 14/12 gigahertz (Ku-band).[81]

NASA also proposed to test three other advanced technologies on this satellite. One was time-division-multiple-access, whereby a ground station would communicate with its satellite only during an assigned period of time. It would then relinquish the connection to another ground station, which had been assigned the succeeding time slot and so on, until the time came around for the original station to resume communication. This kind of multiple access contrasted with the frequency-division-multiple-access then in use, whereby all the stations communicated simultaneously with the satellite, but each used a different portion of the satellite's range of frequencies.

Second, instead of projecting a single radio beam that covered the continental United States, the antennas of the NASA satellite would generate multiple beams, each covering a handful of states. The point of this innovation, as with time-division-multiple-access, was to make better use of the band of frequencies that the satellite commanded. The same frequency range could be used in each of the beams, or, in the jargon of satellite engineering, the frequency could be reused as many times as there were beams. Finally, multiple beams dictated a third technology that NASA proposed to develop with its 30/20 gigahertz satellite. This was an onboard system that could switch signals from one beam to another.

Proponents of the 30/20 gigahertz satellite argued that the research was needed because the number of satellites was growing so rapidly that there would soon be more of them than the C and Ku frequency bands could accommodate. They maintained that U.S. industry could not itself finance the research because the industry was divided into groups with differing interests. On the one hand were the satellite manufacturers, who had an interest in producing advanced Ka-band satellites, but lacked the funds for this kind of high-risk, long-term project, because their profit

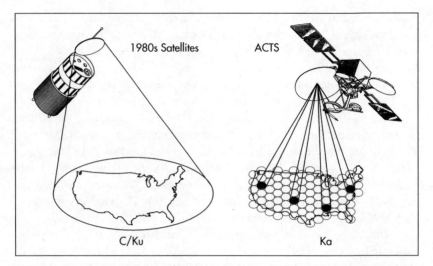

A comparison of satellite coverage. While the antennae of 1980s communications satellites produced beams that encompassed all of the continental United States, the NASA satellite was designed to have several narrow beams, each able to hop from one area to another.
Source: Courtesy of Advanced Communications Technology Co. and the NASA Lewis Research Center.

margins were low. On the other hand were the satellite operators, like Western Union, Intelsat, and AT&T, and the satellite users, like the television broadcasters. They had more profits, but fewer incentives, because when the chips were down, they would have the option of going to foreign manufacturers for Ka-band satellites. Proponents pointed out that this was precisely the reason the NASA 30/20 Ka-band satellite was needed. Japan had already lofted a 30/20 gigahertz test and evaluation satellite in 1977, and was preparing to put up another in 1983. Europe was planning to put up one of its own in 1985. If U.S. satellite manufacturers were to hang on to the dominant share of the world market, it was important that government do research to keep them on technology's leading edge. Such high-risk, expensive research was exactly the proper work of government in its partnership with U.S. industry.[82]

Opponents argued that federal budgets were tight. Congress asked whether, if this research were as vital to the industry as it was represented to be, the companies could not finance it themselves. "Let's think the unthinkable and say . . . the Federal Government is going to get out of it. What is the industry going to do then? . . . Could you not command the capital necessary?"[83] Of course, were the government to elect to do the work and spend the money, industry would naturally accept, but would this not be "corporate welfare?"

Opponents claimed that government and business differed so fundamentally in their approaches to high-risk research as to make federal agencies constitutionally unsuitable for carrying out commercial research. Government mounted large projects because it could muster constituencies for them and could get them through the legislative process. Business, in contrast, funded small projects, which it monitored tightly and terminated promptly if they looked unpromising.[84] Opponents further maintained that government agencies were too far from market demands. As a result, government programs were shaped by engineers who were enticed by what was technologically interesting. Such programs, far from serving industry, were mere "playpens" for federal scientists.

In making the contrary case for the NASA program, proponents invoked a number of examples in which government had done research for the private sector. Most frequently, the example was the aeronautical research that NASA and its predecessor agency, the National Advisory Committee for Aeronautics, had carried out on behalf of aircraft manufacturers. Sometimes the precedent cited was NASA's work on Syncom, or the agency's communications satellite researches taken collectively. Occasionally, someone with a sense of history brought up the Department of Agriculture and its long tradition of research for U.S. farmers.

Indeed, the accusations of "playpen" research leveled against the 30/20 gigahertz project stand revealed as simplistic as soon as we look more deeply at the NACA and Department of Agriculture cases that were so often cited as precedents. This is because historians of these agencies have convincingly shown that "playpen" research is a factor in every program. Every research project is the product of a contentious, if not always explicit, negotiation involving the desire of researchers to do engineering (or science) at the frontiers of their discipline, the needs of the institution in which the researchers work, and the demands of the funders or users of the project.[85] The outcome of the negotiation depends on the power of the several protagonists and on the details of their interaction. If the 30/20 gigahertz program resembled NACA or Department of Agriculture research in certain respects, however, there was one crucial way in which it differed. NACA and the Department of Agriculture had done their work in-house: NASA, in contrast, contracted its research out. This meant that although a project like the communications satellite program was intended to benefit all of U.S. industry, it would begin by benefiting a particular company—the one that held the contract.[86]

NASA also took pains to construct arrangements that would keep it in touch with market needs. In 1977, the agency had surveyed more than 100 companies for advice on setting up the communications project. In 1978, it created an Ad Hoc Subcommittee on Space Communications Applications within the NASA Advisory Council, made up of representatives from satellite manufacturers (RCA and Ford Aerospace), carriers (AT&T, Comsat Corporation, American Satellite Corporation, Western Union, and Digital Communications Corporation), universities (Stanford), and public service groups (the Public Service Satellite Consortium). The subcommittee worked with the Communications Directorate within NASA Headquarters' Office of Space and Terrestrial Applications, providing the directorate with guidelines, and reviewing its plans.

Lewis Research Center also set up a committee of industry advisors, the Carrier Working Group, with representatives from an array of satellite operators and equipment manufacturers, including AT&T, Hughes, RCA, Comsat, and Western Union.[87] Explained one member of the Ad Hoc Subcommittee, "the best assurance that the NASA R&D program in space communications is properly directed towards practical applications is the continuation of the industry liaison with both hardware suppliers and carriers that has been instituted by the Communications Directorate and Lewis Research Center."[88] In fact, these committees were heavily involved in suggesting the technology for the program. But committees so constituted were in no position to oppose the project as a whole. This was true even though some of the participating companies opposed it and placed representatives on the committees chiefly in order to keep apprised of NASA activities. On the contrary, the very existence of such committees helped confer prestige and legitimacy on the 30/20 gigahertz project, while individual members could often be counted on to lobby for it. The industry committees served at the same time as advisors to the program and as promoters of it.[89]

Industry was divided on the program, and the stance of individual firms often reflected, at least in part, their business interests. Comsat, which had recently lost its job as manager for Intelsat and with it Intelsat funding for its laboratory, testified to Congress that satellite operators needed the research and could not fund it themselves. Satellite Business Services, a new joint venture organized to provide links among the dispersed facilities of large business firms, listed Ka-band development as one item in a menu of research areas that were important to their company, but too expensive for them to undertake. Western Union, which had put up its first domestic satellite in 1974 and by 1981 had three *Westars* in orbit, and RCA, which was both a carrier and a manufacturer of satellites, spoke in the program's favor.[90]

AT&T Bell Laboratories, on the other hand, testified against the 30/20 gigahertz project. Its director of communications research, Arno A. Penzias, objected on several grounds. One was that industrial research needed to be tightly tied to development and manufacturing, which could not happen in the government. Another was that the problem the 30/20 gigahertz program was designed to meet, the overuse of the C and Ku bands, could be solved in other ways. Terrestrial fiber optics links could substitute for satellites for links among major cities where there was a high density of traffic. It was also possible to decrease the amount of information that had to be communicated and therefore the amount of frequency band that had to be used by using smart receivers fitted with computers that could reconstruct entire messages from the merest of hints.[91]

The most severe criticism AT&T leveled against the program was "rain fade." Bell Telephone Laboratories had done experiments in the 1970s to see how rain affected 20 gigahertz signals transmitted from satellites, and got results that demonstrated severe attenuation of the signal. Telephone customers demand 99.99 percent availability of their lines, but rain fade at 20 gigahertz restricted availability to 98 percent. Ka-band, held AT&T, was simply not a useful commercial technology.[92]

Hughes Aircraft Company also opposed the project. Hughes had explored 30/20 gigahertz technology in the 1970s and thought it knew more about this frequency range than other companies. It was not keen to see rivals gain comparable expertise through government contracts. Hughes saw the point of NASA doing general research on advanced communications technology, but was adamantly opposed to NASA's placing a satellite in orbit. The point of an orbiting satellite would be to work out problems of satellite operation, and that was a job for private firms, not for government. Hughes derided as nonsense the NASA claim that satellite manufacturers did not have the finances for advanced research. Like AT&T in the Syncom days, but in marked contrast to its own position then, Hughes affirmed that research leading to profitable commercial innovations should be left to the commercial sector.[93]

When it took power in 1981, the Reagan administration also opposed the program. The new administration believed as a matter of principle that federal agencies should limit their development work to technologies for which government was the primary customer. Applied research in support of private industry should be left to the commercial sector.[94]

The White House also argued that NASA's communications satellite program duplicated work that was going to be carried out under the auspices of the Department of Defense. This was the Military Strategic-Tactical and Relay Satellite (Milstar) program for a communications satellite that could transmit messages during a nuclear war. It was part of a plan for strengthening the military's ability to command and control its units that the new Defense Department Secretary, Caspar Weinberger, put at the heart of his Strategic Modernization Program. Milstar shared some features with NASA's 30/20 gigahertz satellite. It had the same downlink frequency of 20 gigahertz, although it used an uplink frequency of 44 gigahertz. It incorporated onboard switching of signals, and it made use of multiple beams, although with the aim of fooling the enemy, rather than getting greater use out of the Ka-frequency band.[95]

In late 1981, OMB deleted from NASA's budget twenty-nine million dollars that the agency had requested for a formal start for the 30/20 gigahertz project in fiscal year 1983. Congress reversed this administration action and voted twenty million dollars for it. But OMB opposition led NASA to restructure the program in the spring of 1982. It renamed it the Advanced Communications Technology Satellite (ACTS), and redefined the relation that was to obtain between it and the ACTS contractor. NASA specified that the contractor for ACTS would be allowed to choose the spacecraft on which the ACTS electronic package would fly. The contracting company would further be permitted to include on the same spacecraft its own revenue-earning communications equipment. NASA expected that this provision would reduce the amount the performing company would seek to write into the contract in the form of a charge for the spacecraft. Further, though the ACTS payload would be designed for a life of ten years, NASA proposed to conduct its own experiments for a period of two years only. After that, the agency would turn the payload over to the contractor for whatever use, be it commercial or experimental, that the contractor wanted to make of it. Finally, the ACTS spacecraft would be given a free launch on the shuttle. On the strength of these changes, which NASA projected would add up to a cheaper project, the administration put ACTS back into the budget, albeit for a minimal amount, for fiscal year 1984.[96]

Milstar, in contrast, sailed through the appropriations process. The administration asked for $79.8 million. Congress, which was as enthusiastic as the Pentagon about upgrading military command and control, provided $142.8 million at the end of 1982.[97] The Air Force, which was to manage Milstar, let two contracts, one for the spacecraft, and another for the electronics payload: The spacecraft contractor would also serve as prime contractor. Lockheed won the prime contract in a competition with TRW. Ford Aerospace, General Electric, TRW, and Hughes bid as a single Hughes-led team and won the payload.

In March 1983, Lewis Research Center sent out a request for proposals for ACTS. Lewis thought that Hughes, Ford, and RCA would all be interested. These were the dominant manufacturers of commercial satellites at the time. Hughes Aircraft Company was "by far the biggest." Ford had inherited the satellite business when it bought the Philco Corporation in 1961. Its satellite-making subsidiary was called Philco-Ford Corporation until the mid-1970s, then, briefly, Aeronutronic Ford Corporation, and finally, since the late 1970s, Ford Aerospace and Communications Corporation. RCA Astro-Electronics built weather and navigation satellites for the U.S. government. It also built communications satellites for the U.S. domestic market, selling most of them to its parent company's satellite operator, RCA American Communications Company (RCA Americom).[98]

The responses were disappointing. Hughes asked AT&T if Hughes might modify one of the Telstar satellites it was building for the telephone giant to carry the ACTS payload. AT&T was unenthusiastic about ACTS and declined. Lacking AT&T's cooperation, and with the Milstar contract in its pocket, Hughes decided not to bid on ACTS. Nor did Ford Aerospace respond to Lewis Research Center's request. The only company that did bid was RCA Astro-Electronics. It proposed to use its own SATCOM vehicle for the spacecraft, and to dedicate it completely to ACTS. Because RCA would not fly any hardware of its own on the spacecraft, this threatened to wipe out the savings NASA would have garnered from using a shared vehicle. Nevertheless, NASA had a bid and the relevant question at this point was whether the company was acceptable. RCA proposed to subcontract the electronics payload to TRW, and the ground systems to Comsat Corporation.[99]

Hughes, which then dominated both the domestic and the international satellite industry, reacted aggressively. It filed a proposal with the Federal Communications Commission in December 1983, stating that it intended to build a Ka-band satellite with its own funds. This satellite would not have all of ACTS features. It would not have onboard switching, for example, nor ACTS's capacity to move its beams from one ground station to another. Nevertheless, the Hughes filing sufficed to strengthen OMB and administration opposition to ACTS on the grounds that the private sector was acting to fund the research.

At this point, then, the ACTS program had been converted into an agonistic field on which Hughes and RCA did battle. Hughes argued that ACTS duplicated Milstar and that technology that was genuinely commercial was bound to be developed by the private sector with private funds. It labeled arguments that RCA had laid before the U.S. Federal Communications Commission in the case as "a transparent appeal by

RCA to have NASA fund its Ka-band system with government money."
RCA maintained that Hughes' satellite "artfully employs all the technical
jargon of ACTS and develops none of the technology."[100] To abandon
ACTS in the face of the Hughes proposal, said RCA, would be to allow
a single company to determine the shape of commercial satellite technology for the entire industry.

The episode ended in mid-1984, when Congress voted ACTS forty
million dollars for fiscal year 1985. NASA and RCA signed their contract
and ACTS was launched. Hughes never took steps to build the Ka-band
satellite it had filed for. The move had simply been an attempt to forestall
a program that its leaders thought unwise in itself and disadvantageous
to the company.

The ACTS controversy served to reveal the fight over NASA's role in
commercially oriented research for what it was: a rich gumbo of contemporary slogans, economic theories, ideology, and self-interest. Among the
1980s' slogans were the bugaboo of international competition and the
"need" to rein in federal spending on civilian projects. Among the economic issues were questions like whether the marketplace or the government is better suited to direct the advance of commercial technology and
whether private industry can be trusted to finance long-term, high-risk research. Among the ideological principles was Reagan's belief that government should limit itself to basic research or research for which it itself
was the customer. In the category of self-interest were Hughes' and RCA's
actions. They graphically illustrated the dog-eat-dog fight for competitive
advantage which characterizes the U.S. aerospace industry.

Conclusion

What impact did the Reagan administration's belief that most economic
activity should be left to the private sector have on the commercialization
of space during Reagan's first term? The answer differs depending on which
part of space we consider. Reagan's philosophy was significant for pulling
new entrepreneurs into the space business. These men and women shared
the president's conviction that private industry can do things better than
government can, and expected that his pro-business administration would
help them. The aid they wanted included things like streamlined regulations for obtaining launching licenses. But private companies—both new
and established ones—did not hesitate to call for direct government assistance. Government funded R&D was one kind of help that private industry asked for. Guaranteed government business and low prices for the use
of government facilities were others.

Unfortunately for them and NASA, the Reagan government's position was that R&D was also better left in the hands of the private sector. Government, the Reaganites maintained, had a natural interest in paying for research on systems for government use, and it had an obligation to pay for generic research that would go into the public domain, but research for revenue-earning commercial technology should be left to industry.

This became an issue in NASA's research into communications satellite technology, the program that after spring 1982 became known as ACTS. Here the Reagan philosophy acted to slow down the program; Congress passed each annual appropriation against administration resistance. Does that mean that in communications, Reagan held back space commercialization? The answer depends on one's view on how useful the ACTS program was to the communications satellite industry.

In fact, the ACTS program has been like those silent movie heroines who face one peril after another, but land on their feet at the film's end. ACTS managers survived the Hughes filing in 1983 and 1984 only to find, by 1986, that one of the chief roles projected for ACTS—the transmission of voice messages between cities with high rates of traffic—was being usurped by fiber optics.

Lewis's comparative lack of experience in communications satellite engineering and the lackadaisical performance of TRW, the electronics payload subcontractor, also took a toll. Added to this was the continual feuding between the administration and Congress over the continuance of the program. There were other problems. ACTS suffered mounting costs. Launching, moreover, which in 1981 had been projected for 1987, was delayed year by year. Nor could the ACTS program get private industry to cost-share by funding the ground stations, which had been one of the objectives from the start. An irate Congress in 1988 imposed a spending cap on the program and forced a reorganization. NASA fired TRW. GE Astrospace, which had taken over RCA Astro-Electronics and with it the ACTS program, took the payload contract in-house.[101]

ACTS was finally launched in September 1993. At first, commercial firms continued to be indifferent and the program was reduced to finding partners among the military. But ACTS promoters had been correct and AT&T had been wrong in seeing the saturation of frequencies as an impending problem. In addition, communications satellite operators were beginning to struggle for a place for their own technology alongside fiber optics in the "information superhighway" that was to link the globe. The high frequency of the Ka band, which led to the rain fade that impeded conversations, raised problems for its use in telephony, but made it eminently appropriate for the transmission of data between distant computers. The fact that ACTS had onboard switching meant that it could be reconfigured rapidly to change the points between which messages were transmitted. This, together with the circumstance that satellites could reach terminals erected at sites where cables had not yet been laid gave it suppleness, and made ACTS-like satellites peculiarly suited to supplement fiber optics links for computers.[102] ACTS was pioneering both a frequency that could relieve overcrowding and a way for communications satellites to position themselves as a flexible and necessary supplement to terrestrial fiber cables. By 1996, ACTS leaders could boast that nearly twenty firms in the United States alone had filed for licenses to build Ka-band satellites.[103]

To discuss the Reagan government's impact on the commercialization of launch vehicles, we need to recognize one fact: launching commercial and foreign satellites was, until the early 1980s, essentially a commercial activity run by a government-industry combine. On paper, NASA bought the hardware and ran the business. But in reality NASA had a hand in how the hardware was designed and produced, while industry participated in the marketing and, under contract to the agency, provided the actual launch services. Thus, business functions—design, manufacture, marketing, and so forth—were distributed between government and industry in a fairly complex way.

During the 1970s, NASA made plans to unilaterally terminate this business. Or, from another perspective, it made plans to buy a different kind of hardware, the shuttle, from a different configuration of aerospace firms, Martin Marietta, Thiokol, and Rockwell instead of McDonnell Douglas and General Dynamics. The ingredient that the Reagan administration brought in was a plan to allow industry to take over the old NASA-industry expendable launch vehicle business and run it as a private venture. It was not the most promising business prospect. Unkind critics even characterized the moves toward a commercial launch industry as mere window dressing. For they knew that NASA was determined to compete with a private launch vehicle industry in every way it could. The *Challenger* disaster, in Reagan's second term, would rearrange the pieces, and invigorate the commercial launch industry.

6

In the Wake of the *Challenger*

ON JANUARY 28, 1986, THE SHUTTLE *CHALLENGER* blew up seventy-three seconds after launch, killing all seven crew members. Any accident that leads to the death of astronauts is a special agony for NASA. In this case, the emotional and political pain was compounded by the presence of Christa McAuliffe as a member of the crew. McAuliffe, a New Hampshire high school teacher, had been played up by the agency and the media. She was to have been the teacher-in-space, the ordinary citizen whose inclusion signified that the shuttle was now a dependable vehicle. Instead, her death pointed up the risks of spaceflight and deepened the effect the accident had on NASA.

The *Challenger*'s explosion also had an effect on NASA's dealings with industry. Not surprisingly the area most profoundly affected was space transportation. As a result of the accident, NASA's relation to the expendable launch vehicle industry was turned on its head; from being that industry's competitor, the agency was transformed into its customer.

The effect of the accident on the space station program was not as great. The mix of public and private elements in the program changed, but the explosion of the *Challenger* figured as only one of a number of causative factors. Nor did the mix change monotonically, with more and more of the space station program moving into the hands of the private sector. Rather, there was movement in both directions. Industry became a little more involved in systems integration. It was a slight difference but it presaged a greater shift down the road. But the idea, accepted by station partisans before the accident, that privately owned platforms would orbit alongside the government facility was defeated by them afterwards, for they came to see such platforms as rivals to the station.

Space Transportation

The *Challenger* accident had two immediate impacts on NASA activities in space transportation. First, it led the administration to ground the fleet of shuttles that was NASA's ferry into space. In early 1986 no one knew how long the shutdown would last; estimates varied from six months to two years. Second, it changed the balance of power between NASA and the shuttle's enemies within the Reagan administration. These enemies included first the Department of Transportation, whose Office of Commercial Space Transportation was just gearing up to license commercial launches, and which viewed itself as champion and spokesman for the commercial launch industry. Second, there was the Department of Commerce, where free-enterprise ideologues advocated turning over space infrastructure, including space transportation, to the private sector to the maximum extent possible. Finally, there was the Department of Defense, whose leaders had been battling NASA for several years for the right to send some of their payloads into orbit on expendable rockets. In 1984, the Defense Department had succeeded in winning the right to buy ten launchers for use over five years and in 1985 had chosen Martin Marietta's Titan over General Dynamics' Atlas-Centaur and NASA's SRB-X. But the struggle had not augmented its affection for the shuttle.[1]

NASA had been weak even before the *Challenger* blew up. Deputy Administrator Hans Mark resigned in late 1984 and for a year, Administrator Beggs resisted the White House's attempts to appoint William R. Graham to succeed him. Beggs thought Graham lacked experience in technology and management. Then, shortly after Graham finally took the post in late 1985, Beggs was indicted on charges connected with a contract at General Dynamics, where he had been an executive prior to coming to NASA. By December 1985, NASA was functioning with only an acting administrator, William Graham, who was new to the agency and lacked support among NASA's managers.[2] By subjecting NASA to an intensely public scrutiny and a barrage of criticism, the *Challenger* accident left the agency still more debilitated.

THE SUSPENSION of shuttle flights impelled the three big manufacturers to take a fresh look at the market for launch vehicles. At this point, Martin Marietta was developing an upgraded version of its Titan (designated Titan IV) for its Air Force contract. It had newly received a separate Air Force contract to convert some old Titan II missiles into launchers. It was also thinking of converting another of its old military launchers into a commercial "Titan III." McDonnell Douglas, which had brought production of its Deltas to an end in December 1984, had turned the factory space over to other products. Delta production equipment was rusting in company sheds and courtyards. Only four complete Delta rockets remained in government inventories, although spare parts were available that could be assembled in case of need. General Dynamics had been trying for some time to work out arrangements to produce commercial Atlas-Centaurs, but none was yet in production.[3]

There were four aspects to the market as these firms saw it. One was the payloads that were to have flown on the shuttle and now needed other transport. The backlog was all the greater because the shuttle office had planned a heavy schedule for 1986. Although the nine times that the shuttle had flown in 1985 were the most to that year, NASA had been planning fifteen flights for 1986. If the shuttles were grounded for six months, NASA estimated there were fifteen to twenty payloads that could never be rescheduled on shuttles; if the shutdown lasted more than a year, at least twenty-five to thirty payloads would go unlaunched.[4]

A second aspect to the market was estimating future demand for launchers of commercial satellites. Vital here was the question of whether the shuttles would continue to book commercial and foreign government payloads, once they resumed flying. But the trajectory that communications technology would take was also an issue. The market for communications seemed certain to increase; the question was, what technologies would service these new communication needs? NASA projections for the number of communication satellites that would be launched in future years had consistently been high. But were they accurate? In 1986, terrestrial fiber optics links were beginning to threaten communications satellites as a means of providing telephone service between major cities. Moreover, an excess capacity of about 35 percent had developed as operating companies had launched larger satellites with more channels. Owners of satellite systems had drawn back from fielding new satellites as a consequence and this translated into a smaller market for launchers.[5]

A third aspect to the market was the Department of Defense. The Air Force had reacted to the shuttles' grounding by ordering extra Titans from Martin Marietta in addition to the ten it had signed up for. It also instituted a competition for twelve smaller boosters. These would be used to loft the Global Positioning Satellites, a series of navigational satellites that had been stranded by the halt in shuttle flights. There were plenty of reasons to believe the Defense Department might procure more expendable launchers, now that it was liberated from its commitment to the shuttle.

The Air Force deliberately used military procurement to promote a civilian launch industry. Into its request for proposals for its twelve medium vehicles it incorporated the requirement that the winning company must offer a variant of its military booster to the commercial market. There were several ways in which furthering a commercial industry served military interests. Requiring its contractor to produce commercial as well as military units would lower the costs of each booster to the Defense Department, as fixed production costs could be spread over a larger number of units. A civilian industry would create excess capacity in case the DoD ever needed to increase its stock of boosters suddenly. And it would save the department the embarrassment of denying civilian satellites passage to space, were it to exercise its prerogative of bumping civilian payloads from the shuttle.[6]

Finally, the aerospace companies had to give thought to the competition for commercial cargoes, such as the already operating European Ariane rockets. Down the road they would also face the Chinese, who were signaling their desire to put their Long March boosters in the commercial market. The Soviets wanted to offer their Protons. The Japanese had commercial launchers in the development phase. The short-term commercial market, roughly from 1988 through the early 1990s, with its demand from backlog spacecraft looked quite good. But would the longer-term market support more than one or, at most, two U.S. firms?[7]

The strategies these considerations prompted varied with the company. McDonnell Douglas wanted the administration to continue flying commercial payloads on the shuttle as it had invested a considerable amount in its Payload Assist Module, which was designed to send spacecraft from the shuttle into other orbits. Martin Marietta and General Dynamics, to the contrary, lobbied first and foremost for an administration pledge to ban commercial cargoes from the shuttle.[8]

The issue of whether to permit the shuttle to carry commercial payloads was debated in the president's Senior Interagency Group on Space, where NASA was represented, and in his Economic Policy Council, where it was not. Acting Administrator Graham was not opposed to banning commercial cargoes from the shuttle. He also wanted NASA to move toward a mixed launch fleet, composed of the shuttle and expendable vehicles, and he called for NASA to study the option of sending some of its own payloads on commercial launchers. Graham, however, was not destined to stay at NASA's helm. Beggs had struck a deal with the administration that he would resign if the administration would bring back James Fletcher to serve another term as NASA administrator. Fletcher was appointed and took office in May 1986.[9]

Fletcher, reversing Graham, fought hard to keep commercial payloads for the shuttle. By this time, however, there were too many forces stacked against him. In addition to the Departments of Commerce, Transportation, and Defense within the administration, there were outside opponents. They included two of the three launch vehicle manufacturers, influential members of Congress, and members of the scientific community. The latter group was tired of the extra expense and work of "man-rating" its experiments for the shuttle, and chagrined at the delay the *Challenger* explosion had occasioned in the launching of major science spacecraft, like the Hubble telescope. On August 15, the fight was resolved. President Reagan issued a statement specifying that "NASA will no longer be in the business of launching private satellites."[10] The road was now clearer for the big manufacturers.

THE MORE the large launcher manufacturers found the market inviting, the more many of the start-up companies found themselves squeezed. Perhaps the hardest hit was Transpace Carriers, Inc. (TCI). In 1984, the company had signed a preliminary agreement with NASA that included the right to buy back from the agency the stock of surplus Delta rockets and spare parts. But in June 1985, the Strategic Defense Initiative Organization had asked NASA to turn over two of the Delta rockets promised to TCI for the Star Wars program. In August 1985, furthermore, President Reagan had reneged on his promise to establish a price for commercial payloads flown on the shuttle for the period starting with 1988 that would represent the full cost. Instead, fearful that the United States would lose its share of the launch market to the Ariane, Reagan had set the price at about half of what the Department of Commerce and TCI estimated as just.[11] TCI itself was having cash flow problems so severe that it could not pay salaries or the rent on its Greenbelt, Maryland, offices. Nor was it meeting NASA's conditions that it line up three customers and financing that NASA saw as reliable enough to fund Delta production.[12]

TCI nevertheless had pressed on, trying to find ways to lower its asking price for a Delta launch so as to meet the competition of the shuttle. Meanwhile, the company was urging NASA to sign a final agreement. That step, TCI executives thought, would be the "open sesame" to financing and customers. NASA, however, was only willing to grant TCI extensions of the preliminary agreement. By the end of 1985, it had granted five.

The *Challenger* accident of early 1986 at first seemed to create a more favorable situation for TCI. The shuttle was down, and there was reason to hope that when it resumed flights it would be barred from flying commercial payloads, including Delta-sized payloads. But NASA still refused to execute a final agreement until TCI had customers and capital, while TCI still maintained it could not attract customers and financing without a final agreement.

As summer passed into fall, McDonnell Douglas was reconsidering its part of the bargain. Enticed by the carrot of the president's August 1986 proclamation that the shuttle was to leave commercial payloads to private industry, and goaded by the Department of Defense's insistence that any contractor that sold launchers to the military would also have to sell them commercially, McDonnell Douglas, in October, finally announced its intention to produce commercial launchers.[13]

By that time, TCI had secured some financing and developed plans to secure the rest. A sixth extension of its preliminary agreement had ended on May 31, 1986. NASA administrator Fletcher now concluded that TCI had never demonstrated the financial robustness required to succeed in the launch business. The goal, wrote Fletcher, was "reestablish[ing] the Nation's space program in the most effective and expeditious fashion." It was the manufacturers themselves, he maintained, who were "in the best position to quickly establish a viable commercial expendable launch industry."[14] For the Delta, that meant McDonnell Douglas. By 1988, Transpace Carriers was in extremis: the company had gone into bankruptcy and, as a kind of death rattle, was preparing lawsuits against McDonnell Douglas and NASA.

Those small companies that had organized to provide auxiliary services to the shuttle were also in a difficult position. Astrotech Space Operations, for example, had turned a profit in 1985 when it processed ten payloads for the shuttle. With the grounding of the shuttle fleet, its business vanished. Astrotech cut its Titusville, Florida, staff back to three caretakers and began to study adapting its payload processing service to expendable launch vehicles. Its parent company, Astrotech International, had had a brief moment on the public stage in the spring of 1986. A combination of Acting Administrator Graham's enthusiasm for free enterprise and the loss of the *Challenger* had led NASA and Congress to reopen the question of private financing for a shuttle—now to be a replacement for *Challenger*. But the Reagan administration quickly came to the decision that the government would pay for a replacement. From Astrotech Space

Operations' viewpoint, perhaps the most cheering part of the story was that Astrotech International, by this time hemorrhaging money, sold the Space Operations division off in 1987 to a partnership of Westinghouse's Wespace subsidiary and Space Industries, Inc. With this support, Astrotech Space Operations hunkered down to wait for the first commercial launchers to come on line.[15]

Orbital Sciences Corporation had predicated its business on building upper stages for the shuttle. Its first upper stage was to be for a NASA scientific mission, the Mars Observer. But Orbital Sciences also counted on being able to sell its upper stage to some of the shuttle's commercial customers. The grounding of the shuttle fleet dealt a double blow to the company. NASA began to talk about delaying the Mars Observer from its scheduled launch date of 1990 until 1992 at about the same time Reagan decided to eliminate most commercial payloads from the shuttle manifest.

Orbital Sciences went to NASA in early 1987 with a bold plan. It suggested that NASA launch the Mars Observer on an expendable Titan instead of the shuttle. Further, since NASA had not yet decided what kind of vehicles to buy for the expendable launch vehicle component of the "mixed fleet" it had been talking of since the *Challenger* accident (NASA did not even know where the money would come from for the expendable part of a mixed fleet) Orbital Sciences offered to put up the initial deposit and the early progress payments for the Titan. NASA would only have to take over payments and reimburse Orbital Sciences in October 1988. The benefits to the company would be that it would be able to get the Mars Observer up by 1990 instead of 1992 and it could demonstrate to the commercial world the efficacy of its TOS (transfer orbital stage), suitably modified from shuttle-compatible to Titan-compatible. NASA rejected this proposal. It did award Orbital Sciences a second TOS contract, this one for the ACTS launching. But the ACTS program was going slowly; even its continuance was in doubt. Orbital Sciences' position was unenviable.[16]

Fortunately for Orbital Sciences, a new market was starting to appear for small launchers. Officials of companies like Space Services had been predicting the emergence of such a market for a half decade. No one, perhaps, had been more optimistic in this regard than George A. Koopman, president of American Rocket Company (AMROC). AMROC had been organized by some of Starstruck's leaders and employees in 1985, after Starstruck's president, Michael Scott, had peremptorily dissolved his firm in 1984. In 1986, Koopman predicted there would be "hundreds of payloads" available to a small-launcher industry before 1992. Many of these, he expected, would be commercial. In 1987, a commercial market for small launchers was nowhere to be seen, but a government market was quite suddenly visible. The Defense Advanced Research Projects Agency (DARPA) began to need them.[17]

DARPA's idea was to augment the ever more costly and complex military satellites then being fielded with systems of low-weight, single-purpose, cheap satellites. "We're looking to cultivate a Bic pen or a Pampers philosophy," a DARPA spokesman told the press, "use it once and throw it away."[18] Light satellites, in the fifty to six hundred pound class, were becoming more capable as better sensors were invented, electronic circuits were made even smaller, and small, high powered, space-qualified computers were developed.[19]

Light satellites need small launchers. Orbital Sciences decided to take up this challenge. Already, company scientists had floated the notion of launching small boosters from military bombers. Working with solid rocket manufacturer Hercules Corporation, Orbital Sciences put together a proposal for DARPA based on bomber launchings. Launching from a site that was mobile, and thus not as vulnerable to enemy attack as a fixed launch site, was a powerful attraction for military planners. In May 1988, DARPA sent Orbital Sciences-Hercules a letter of intent, accepting their proposition. Orbital Sciences had begun its history by targeting the civilian and commercial user—NASA scientific satellites and commercial users of the shuttle. Even as it proposed the new small launcher—dubbed Pegasus—to DARPA, it still had its eye on commercial customers.[20] But the shape of the market was a military one and the company found it necessary, at the least, to take a military detour.

Other of the small launch companies began to find in the military's new fancy for "lightsats" a possibility for earnings. Space Services, Inc., got a DARPA contract for engineering studies on the Conestoga rocket in the beginning of 1988. The Army used Strategic Defense Initiative funds to give Gary Hudson's Pacific American Launch Systems, a firm he had organized after the Percheron debacle, a $400,000 contract in summer 1988. American Rocket Company nearly obtained a similar contract directly from the Strategic Defense Initiative Organization eight months earlier. Only its desperate financial situation had prevented this contract from being consummated.[21] A fresh wind had finally arrived to fill the sagging sails of the small booster builders.

THE AIR Force chose McDonnell Douglas to provide the launch vehicles for the global positioning satellites at the end of 1986. The company had offered an upgrade of its Delta rocket, called Delta II. By that time McDonnell Douglas had lined up four firm commercial orders and an Air

Force official pronounced its ability "to attract commercial launch business . . . a key factor in the decision."[22] At this point, Martin Marietta and McDonnell Douglas both had Air Force contracts. General Dynamics had none, but proclaimed its intention to field commercial launchers anyway.

Air Force Secretary Edward C. Aldridge returned to Congress in the fall of 1987 for funds to run yet another expendable launch vehicle competition. The request raised some eyebrows: "There is some question of whether or not the Air Force really has as one of its major goals the benefit of subsidized commercial industry," remarked Senator Cohen. "Absolutely not," replied Aldridge. The Air Force goal was purely that of finding the best system to place the Defense Satellite Communication System in orbit.[23] The competition this time was between a McDonnell Douglas-Martin Marietta team and General Dynamics (GD). General Dynamics spokesmen did not hesitate to play the commercial card on their company's behalf: "Said one GD source: . . . 'Whichever company wins the . . . competition will have strong appeal for commercial customers'" General Dynamics also laid emphasis on the importance to the military of retaining multiple sources. "GD is pitching the Air Force with the concept of keeping three launch lines going and so enhancing launch resiliency."[24]

General Dynamics won the contract in May 1988 with Atlas II, an upgrade of the Atlas that it bound itself to develop with its own resources. Now all three big manufacturers were in the game with plans to sell commercial launchers. McDonnell Douglas and General Dynamics were going to offer variants of the Delta II and Atlas II they were making for the Air Force. Martin Marietta was going to convert its Titan 34D, then being used by the Air Force as the service waited for the new Titan IV to come on line, into a commercial "Titan III." "In fact," Craig Russell Reed has written in his study of government and the U.S. launch industry, "in the cases of each of the three large launch vehicle manufacturers . . . U.S. government procurements provided the base upon which the company assumed its viability as a commercial launch service provider."[25]

If the Air Force was enthusiastic about restoring the production lines of the big three launcher manufacturers and encouraging them to branch out into commercial launchers, however, it was far from enthusiastic about another, more radical idea. This was to have the government purchase launch *services* instead of launch *vehicles*. When the government purchases a piece of hardware the equipment must conform to government specifications. Government monitors pass on its compliance and industry employees are hired to work with these monitors and to prepare documentation for them. When a company proposes an improvement in a piece of equipment, the change must first be translated into a change in the specifications. On the other hand, when the government purchases services, like the placing of a satellite into a particular orbit on a particular date, the government no longer has oversight over the hardware the supplier chooses to use. Supporters of the idea of moving from purchasing launchers to purchasing launch services therefore argued that the change would save money—up to 25 percent—that would otherwise be devoted to government monitors and extra company employees or extra paper work. They also maintained that the change would speed innovation, as new ideas would no longer have to travel the arduous route into new specifications.[26]

The Air Force opposed the change.[27] It argued that a launch failure of a military payload could cost the military not only the price of a satellite and booster but a capability that might be vital to a battle.[28] It pointed out that it might need extra features in its launchers that would not be standard in commercial vehicles. In battling against the shuttle-only policy before 1986, the Air Force had, in part, been fighting to keep access to space within its hands. Now it did not want to lose control to private firms.[29]

NASA was no more eager than the Air Force to relinquish its right to oversee the manufacture and operation of the vehicles that industry produced to send up its cargoes. The agency could not argue national security, but others of its reasonings echoed Air Force arguments. NASA pointed out that its spacecraft were often one-of-a-kind so that a loss would be catastrophic. Its cargoes were also far more expensive than commercial spacecraft. (Typically, the cost per pound of NASA and DoD spacecraft were tens, and even hundreds, of times higher than commercial communications satellites.)[30] NASA also explained that the web of regulations that surrounded federal procurements made it difficult to execute the kind of launch contracts that the private sector utilized.[31] Critics of NASA's position charged that it was at bottom a question of turf. Overseeing the manufacture and operation of launch vehicles was a substantial program for NASA, as it was for the Air Force. Budgets and teams of personnel were involved.[32]

In 1987, NASA locked horns directly with the Department of Commerce over the GOES (Geosynchronous Operational Environmental Satellite) weather satellites. The GOES craft were operated by the National Oceanic and Atmospheric Administration, which was housed in the Commerce Department. NASA, however, was traditionally responsible for launching them. Bumped off the shuttle by the *Challenger* accident, the satellites needed an expendable launch vehicle. But Commerce wanted to use a commercial contract for services, while NASA wanted to procure launchers. At the Commerce Department Gregg R. Fawkes, whom Commerce Secretary Malcolm Baldridge had appointed to head an Office of Commercial Space Programs in early 1987, was in the forefront of the push to purchase services. Fawkes, who had received his initiation as a member of Klaus Heiss's staff, believed that the infrastructure for space—the transportation systems, the space stations, the man-tended space platforms—should be provided to the maximum possible extent by industry.[33] Partisans of this view thought the private sector could furnish infrastructure more cheaply than the government, and that a cheap infrastructure would, in turn, stimulate companies to enter the space industry.[34] Commerce threatened to stop using NASA as its procurement agent. NASA planned to retaliate by refusing to give the Commerce Department technical aid in evaluating launchers.

It was not until the fall of 1987 that NASA acquiesced to a "commercial" contract for launch services and Commerce agreed to retain NASA as purchasing agent. NASA's "commercial" contract with General Dynamics for the Atlas I, signed in 1988, contained provisions for considerably stronger oversight than was usual in such documents.[35] Nevertheless it was a turning point in NASA's methods of handling launchers.

In February 1988, the Reagan administration issued a new pronouncement on space that included a "Commercial Space Initiative" heavily influenced by members of the Commerce Department.[36] Among the provisions of the new policy statement was a restatement of the requirement that federal agencies buy private sector launch services rather than procuring launch vehicles and conducting the launch operations themselves, to the fullest extent feasible.[37] NASA now began to take steps to bring its practices into compliance with administration edict. In the fall of 1988 NASA sent out a request for proposals for launch vehicle services in the medium class and in 1989 it initiated a procurement for small launch services. The medium ELV contract was awarded to McDonnell Douglas in 1990 for three firm and one optional Delta launches. The small ELV contract went to Orbital Sciences in 1991 for seven launches. NASA also executed a handful of contracts for the launch of one or another single craft.[38] Critics would continue to charge that the agency violated the spirit of the policy by writing commercial contracts that were full of oversight provisions.[39] Looked at historically, however, three years had brought a remarkable change. In 1985, it was NASA policy to control U.S. transportation into space, but by late 1988, NASA was beginning to buy its space transportation from private industry.

NASA'S ROLE in the rise of a commercial launch industry in the second half of the 1980s had both an inadvertent and a deliberate component. The inadvertent part, the explosion of the *Challenger*, had the greater effect. But the deliberate part has more claim on our attention. How did NASA react to the series of actions that were sparked by the *Challenger* accident and led to an active U.S. launch industry? These actions were (1) the decision to keep most commercial payloads off the shuttle; (2) the decision to procure substantial numbers of expendable launch vehicles for government use; (3) the decision to buy small launchers for "lightsats"; and (4) the decision to stop procuring launch vehicles and start procuring launch services.

In regard to the first action, we have seen that NASA fought it from the time that Fletcher was nominated for his second term as administrator until the time, in the summer of 1986, that the White House reached its decision. Fletcher's motive was the same that had impelled NASA since the 1970s: to keep down the cost of shuttle flights by securing the maximum amount of cargo, and to bring in fees from commercial payloads to offset some of the shuttle's costs. As regards the procurement of expendable launch vehicles, a step that was vital to bringing the Delta, Atlas-Centaur, and Titan back into production, this was the work of the Air Force and not of NASA.

NASA did help McDonnell Douglas into production by clearing Transpace Carriers from its path. To the Congress, the agency offered, as reason, the common good. The United States needed to get its space transportation system back in action as quickly and efficiently as possible, and the launch manufacturers were in a far better position to accomplish this than middlemen like TCI. This was certainly supportable. In addition, a strong argument can be made that TCI was misconceived from the start, that the only function it had been organized to perform was to sell a product under circumstances in which the product's actual manufacturer saw no market. Nor did Grimes and his colleagues show great business skill. Granted six successive extensions of their agreement with NASA they were still not able to line up investors and customers. Nevertheless, it is patent that TCI was also a victim of the *Challenger* accident. NASA was forced to rethink its priorities and to elevate restitution of U.S. launch capacity to highest place. McDonnell Douglas found it attractive to market its own Delta rockets. The result was that TCI was cast aside.

The creation of a market for small launchers was also primarily the work of an organization within the Department of Defense, in this case DARPA. NASA ultimately became a part of the market but cannot be said to have spearheaded it. Finally, it was Congress and groups within the administration that pushed for the transformation from procuring launch vehicles to procuring launch services. Here, the Department of Defense as well as NASA dragged its feet to some degree. Both agencies worried about loss of control and the endangerment of precious cargoes.

In general, we see that NASA acted to delay the emergence of a commercial launch industry in the first half of 1986 and helped push it along later. But in all cases, the momentum it contributed, whether positive or negative, was small.

One striking feature of the emergence of the launch industry was that the overwhelming part of the market was the government—fully 90 percent of the launches were for the Department of Defense or for NASA.[40] In view of that, some have asked whether this "commercialization" was anything more than an alternative form of contracting. The answer seems clear when we compare the way NASA ordered its GOES launches in the late 1980s to the way it ran the expendable launch business in the 1970s. The aerospace companies simply had more control over design, testing, manufacturing, and quality in the 1980s, and less paperwork.[41]

But although commercialization was not identical with contracting, the rise of the U.S. commercial launch industry furnishes additional examples of the connection that exists between being a contractor in space and being a manufacturer of commercial products. Perhaps the outstanding case is that of McDonnell Douglas. It began as the one firm among the big three producers that was least interested in marketing its launcher commercially. It was dragged willy-nilly into greater commercial production by its desire to sell Delta rockets to the Air Force. And it wound up as the company that made the first commercial shot, putting the British Satellite Broadcasting System's Marcopolo I into orbit in August 1989.

The evolution of a commercial launch industry within the circumstance of a largely government market is noticeably different from the emergence of the U.S. semiconductor industry or the laser industry. In the laser and solid state electronics cases, it is true, the new industries took their baby steps under the protective wing of the U.S. military. As time went on, however, commercial demand grew alongside government demand to a point where the former eventually predominated. The emergence of the launch industry was much more a matter of artifice. Indeed, it was as much a case of command economy, or government inducement, as the absence of a launch industry before 1986 was a case of government impediment.[42] Only the impediments had been placed by NASA. In the provision of inducements, NASA was a minor actor, except through the contingent event of the *Challenger* disaster.

Space Stations

When NASA Administrator James Beggs formulated the plans for a space station in 1981 and 1982, he was insistent that NASA serve as its own prime contractor. He had several reasons. First, the station was conceived as an "evolutionary" project: an initial configuration would be put in orbit, with other bits and pieces added over its thirty-year lifetime, as funding permitted, to enhance its capabilities. Beggs did not want NASA to have to rely on a single contractor over the course of such a long project. Second, the station was conceived as an international program, with modules and fixtures to be contributed by Japan, Canada, and Europe. Beggs worried about the problems the United States would have with its foreign partners if it gave the job of prime contractor to a U.S. firm. Finally, a prime contractor would necessarily be responsible for integrating the subsystems and ensuring that changes made to one part did not disrupt the functioning of other parts. Systems integration is among the most highly skilled work in engineering and one of Beggs' aims for the agency was to raise the technical level of NASA personnel. By keeping systems integration within NASA, Beggs hoped to help NASA engineers hone their proficiency.[43] Such a policy would also help attract to NASA the best of the new crops of graduating engineers. These, of course, are the same reasons that industrial companies often seek to secure the integration function for themselves.

At the same time he was establishing the principle that NASA serve as its own prime contractor, Beggs and his space station staff were hammering out a management scheme for the program. Political considerations dictated that as many centers as possible get a piece of the action. That would suggest that the station be divided into relatively independent pieces so that each of the participating centers could work up its part in relative autonomy. But Beggs was promising a low-budget station, only eight billion in 1984 dollars, and one way to economize would be to standardize components common to all pieces and "mass produce" them. It would also be cheaper if some systems could be centralized, like electrical power, communications, and thermal control. All this necessarily meant that the pieces doled out to the centers would be less independent. Each center would have to design its part of the hardware so that it would fit with the centralized systems and use standardized parts. Furthermore, the job of designing the centralized systems would also have to be divided up among the centers.[44] Whatever management team was put in place would have to deal with a complex distribution of tasks among the centers and a messy set of interfaces among the subsystems.

In 1983, NASA chose Johnson Space Center in Houston as the lead center for the station. It placed the program management (which it called Level B) there. Above Level B was the Office of the Space Station (called Level A) at Washington, D.C. headquarters. Below it were project offices at Johnson, Marshall, Goddard, and Lewis, the Level C offices. It was at Level C that industry would enter, because the work at the project offices was to be carried out with the help of contractors.

Lines of responsibility were two-fold. Project managers at Level C reported to Level B management, which in turn reported to Level A in Washington. Project managers also reported to their center directors, who reported directly to the NASA administrator. Level B was given the responsibility for systems engineering and integration.[45] But the unwieldy management and the complicated interfaces between subsystems made the task of integration more difficult.

IN 1982 and 1983 the station was expected to consist of a core, inhabited, structure, accompanied by a number of unmanned platforms. It was not clear whether NASA would advocate a single such system, or two, one circling in a polar orbit, and the other at an inclination of twenty-eight degrees to the equator. The question of ownership was equally fluid at this point. Some of the platforms that were dedicated to commercial uses might be privately owned. Indeed, it was contemplated that some of the modules on the station proper might be financed by private capital.[46] To look more seriously at private-industry modules, NASA contracted with Booz, Allen and Hamilton, Inc., working with the Weinberg Consulting Group, to see if they could assemble a group of venture capitalists willing to finance an astronaut-tended, fee-for-service laboratory attached to the station.[47] There was also talk of a NASA unmanned free-flying platform as a possible fall-back if the station were not approved by the Reagan administration.[48]

In this atmosphere thick with schemes for various kinds of public and privately financed stations and station components, a number of companies came forward with concrete business plans. One was Fairchild Space and Electronics Company. It proposed a platform it called "Leasecraft." Leasecraft would be placed in orbit by the shuttle and retrieved into its cargo bay from time to time to allow old payloads to be returned to earth and new ones installed. Fairchild envisioned renting space to industrial users—McDonnell Douglas' electrophoresis program was one customer it had in mind—and also to NASA itself. Such a platform would have the virtue of accommodating payloads which needed more time in space than the shuttle, with its one to two week trips, could provide.[49]

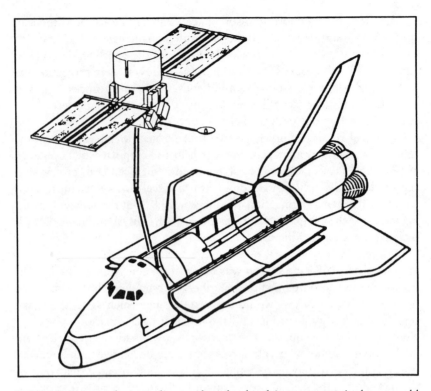

Fairchild's Leasecraft. According to plan, the shuttle's remote manipulator would periodically grapple this free-flying satellite into its bay, exchanging fresh for processed material. The payload went back to earth while Leasecraft continued in orbit.
Source: Courtesy of *Aviation Week & Space Technology.*

Another firm was a start-up company, Space Industries, Inc., headed by Maxime Faget. Faget, who had retired from NASA in 1981, had joined a consulting firm. There he had been approached by two Houston architects and a scion of an oil company interested in fielding a privately financed space station. That plan seemed to Faget overly ambitious. What did make sense to him was a free-flying platform that would be tended from time-to-time by visiting astronauts. It would be a module that could be sent up on a single shuttle launch and that could be expanded, as needed, by the addition of identical modules. As well as sitting in orbit for longer periods, such a space platform would have more power available for experiments than the shuttle did and gravity several orders of magnitude lower. Faget and long-time colleague Caldwell C. Johnson designed the module, which they called the Industrial Space Facility (ISF), and in

1982 he and two others founded Space Industries to promote it. The ISF would not be retrieved into the shuttle's cargo bay, like Leasecraft, but would have a docking port. The shuttle could dock there and by living parasitically off the platform's power system, it would be able to remain for up to twenty-five days. Such a platform, Faget projected, would give the shuttle somewhere to go while waiting for the NASA space station. Once the station were in orbit, the ISF might become one of the private modules attached to it.[50] Like Fairchild, Space Industries looked to McDonnell Douglas and NASA as the most promising customers.

Other companies offered plans that would insert private ventures into the shuttle itself. Spacehab, a Seattle firm organized in 1984, proposed a pressurized module in the shuttle cargo bay that astronauts could enter through a tunnel. The module would have twenty spaces in it to accommodate astronaut-tended experiments. Spacehab would pay NASA for carrying the module, and make its money by renting out the experiment spaces to commercial, university, and government customers. The External Tank Company (Etco), founded in 1985, wanted to make use of the shuttle's external tanks, which were normally jettisoned before the orbiter reached orbit. Etco would have NASA bring the empty tanks all the way to orbit and with the help of the shuttle convert them into crewed laboratories for scientific and commercial experiments.[51]

Beggs and NASA were receptive to these proposals. In particular, the Space Industries platform, scheduled to go into orbit before the station was constructed, could serve to test mechanisms for docking the shuttle, a crucial bit of technology that would ultimately be needed in order to join the shuttle to the station. NASA signed a Joint Endeavor Agreement with Fairchild in 1983 for free shuttle flights so that the company could try out its ideas for Leasecraft. It signed a Space Systems Development Agreement with Space Industries Inc. in 1985, affirming that Space Industries' Industrial Space Facility would be allowed to take the shuttle on a fly-now pay-later basis. In early January 1986, NASA signed a Memorandum of Understanding, essentially an agreement to keep on talking, with Spacehab.[52]

BY THE time the request for proposals for detailed conceptual design (Phase B) of the space station was presented to industry in September 1984, the division of subsystems among the centers had been laid out. Johnson Space Center got the structural framework and the job of integrating into it the station's elements. Marshall was to design the architecture of the modules, including a habitation module, a laboratory module, and resource and logistic modules. Marshall was also put in charge of the systems that would be common to the modules, including the life support systems, needed for both the habitation module and the laboratory module. But it did not get the attitude control and thermal control systems, assigned to Johnson to make the division of work between it and Marshall more even. Nor did it get the outfitting of the habitation module, also given to Johnson in recognition of the fact that the astronauts were based in Houston. Goddard was tasked with outfitting a second laboratory module. Lewis got the power system.[53]

Contracts were awarded in April 1985. Boeing and Martin Marietta were selected as contractors to Marshall Space Flight Center, Rockwell and McDonnell Douglas were to work with Johnson Space Center, General Electric and RCA had the Goddard contracts, and Rocketdyne and TRW those for Lewis Research Center. Work on Phase B began and immediately the defects of the management system began to show themselves. The multiple interfaces among the different tasks and the overlaps among the responsibilities that had been assigned to the centers necessitated constant arbitration. But the lines of responsibility for effecting decisions were not well drawn. The Level C managers could go to Level B at Houston, but they could also take their concerns to their center directors, who reported to the NASA administrator, bypassing the Level A Headquarters Office of the Space Station. At Johnson, the Level C project manager reported to the Level B program manager *and* to the center director, giving the Johnson Space Center director the job of settling the disputes between C and B. To make matters worse, the Level B program manager, Neil Hutchinson, was inexperienced and lacked an aptitude for management, while the head of Level A, Philip Culbertson, was distracted by having to serve simultaneously as NASA acting deputy director. Three years later, Hutchinson would recall, "[The] management difficulties . . . were just awesome. It really took a toll to adjudicate the disputes that arose between the field centers . . . They were constant, they were difficult, they were sometimes vicious, and most of the time they were very parochial."[54]

In September 1985, Philip Culbertson, associate administrator for the Space Station Program, appointed a committee to recommend changes in management and in the distribution of work among the centers. Before the committee could complete its work, however, the *Challenger* exploded. The tragedy brought national attention to NASA's management of the shuttle and by extension it invited outside evaluation of the management of NASA's other programs.

Within a week of the accident, President Reagan appointed a commission headed by former Secretary of State William P. Rogers to investigate it. When James Fletcher returned to NASA in April to again take the reins as administrator, it was clear that the Rogers Commission report, due in June 1986, was going to fault NASA's management of the shuttle. The press was already criticizing NASA management across the board. Fletcher decided to take the initiative. He arranged to have the National Academy of Public Administration set up a study group to review NASA management policies, with Samuel C. Phillips, the man who had directed the Apollo program, as its head. The Phillips committee succeeded in submitting its report to Fletcher shortly after the Rogers Commission did.[55]

In managing the Apollo program, Phillips had centralized control in his Washington, D.C., office. He had hired an outside industrial company, Bellcom, as a kind of "private" source of expertise on systems engineering. Not unexpectedly, in those of its recommendations that touched on the space station, the Phillips committee suggested relocating the program management (Level B) from Houston to the Washington area. It called for NASA to give Level B management a hand with systems engineering by hiring a contractor with that expertise.[56]

These changes, which NASA would implement in early 1987, in retrospect appear as the first step on the path toward greater industry involvement in work that NASA had, initially, determined to keep for itself. A team of companies headed by Grumman would win the program support contract in July 1987, and would function as a kind of aid-of-all-trades to the program management office.[57] The change might have come without the *Challenger* accident, because well before that happened disaffection with the structure of space station management was palpable. But as the events played out, the *Challenger* was the proximate cause.

WHILE THESE changes in station management were proceeding, another effect of the *Challenger* accident was playing out: the diminution of private sector interest in space-based manufacturing. Before the accident, McDonnell Douglas and its Ortho Pharmaceuticals partner had been NASA's star exhibit for such interest. McDonnell Douglas had already flown five electrophoresis experiments on the shuttle's middeck before the end of 1984, using first a piece of equipment that could function twelve hours at a stretch, and then an improved model that gave continuous service for one hundred hours. The company had separated both animal proteins and cells and had demonstrated that it could obtain high purity and output hundreds of times greater than what could be achieved with ground-based electrophoresis. It was building a prototype electrophoresis factory that would be placed in the shuttle cargo bay and, automated to obviate the need for astronaut attendance, would run five days without stopping.[58]

In September 1985, however, Ortho Pharmaceuticals had withdrawn from the first product development of its joint venture with McDonnell Douglas. Responding to the competition of a new bioengineering firm, Amgen, Ortho decided a ground-based bioengineering process for erythropoietin would be cheaper and quicker to prove out. Bioengineered skin cells would replace kidney cells as the raw material's source. The mixture would still have to be purified, but the abundance and cheapness of the new raw material would change the economics for purification. By contrast, the space-based process had run into some bottlenecks: one was the difficulty McDonnell Douglas and Ortho were having in negotiating with NASA a favorable and stable price for the time, starting in 1987 or 1988, when experiment had given way to manufacturing and they would be reimbursing NASA for shuttle flights. They needed a firm price for this service to put into their business plans, and they were having trouble obtaining it.[59]

McDonnell Douglas started talks on partnering with 3M Corporation's Riker Laboratories, and pushed on. It had flown its sixth experiment in the shuttle middeck in April 1985, and it flew another in November. It was scheduled to fly its prototype laboratory in the shuttle cargo bay in July 1986. For the future, it had its eye on Space Industries' Industrial Space Facility. By this time, almost one hundred fifty employees were working on the project.

Then came the *Challenger* accident. McDonnell Douglas could not have been in a worse position. It had lost its pharmaceutical partner. It was in a race with a ground-based technology, a race that found it hobbled by the inability to get back into orbit. In March, 3M Corporation pulled out of the discussions. Theoretically, McDonnell Douglas could have gone on to compete on the ground, because it had found it necessary to do some genetic engineering of its own in connection with the erythropoietin project. But it had been precisely the nexus between pharmacy and space that had brought the company into its Electrophoresis Operation in Space in the first place. As James T. Rose, the program leader, was to explain, "without having the space hat on, from McDonnell Douglas' point of view [we] did not want to get into the pharmaceutical business."[60] By the end of 1986, McDonnell Douglas had terminated the program.[61]

Other commercial materials processing projects were less devastated, but that was only because they were more embryonic. The space industrialization community had already been uneasy over the shuttle as a key element in their plans—the unpredictability of its pricing, the irregularity of its flight schedule, the possibility of getting bumped from the manifest to accommodate Department of Defense hardware.[62] These were things that from the start had made getting capital for start-up companies difficult. Now they confronted something that was worse, the absence of shuttle flights altogether.

At the same time, NASA priorities had changed. For Fletcher and for Dale Myers, whom Fletcher had brought back to NASA as deputy administrator, the central task was to revive the shuttle program and they bent most of their energies to it. Space commercialization had necessarily to take a back seat. Beyond that, Fletcher and Myers wanted to see industry put more of its own, rather than the government's, money behind commercial space ventures. They were suspicious of schemes like the fly-now-pay-later arrangements NASA had been negotiating with some companies. It smelled to them like subsidizing industry.[63]

In June 1987, Myers, against the opposition of the Office of Commercial Programs, consolidated NASA's microgravity programs within the Office of Space Science and Applications. Simultaneously, he acted to reduce the Office of Commercial Program's role in setting expendable launch vehicle policy and to make it more difficult for firms to obtain free rides for precommercial experiments. In August, Gillam retired from NASA and his position as head of the Office of Commercial Programs, leaving behind (according to *Aviation Week & Space Technology*) a demoralized staff.[64] With the flagship McDonnell Douglas project dead, with venture capital firms more skeptical than ever, with transportation into orbit suspended, and with the diminished stature of the NASA office that advocated for space commerce, it is hardly surprising that the private sector began to lose interest.

But it was precisely the space industrial community, and above all McDonnell Douglas, that companies like Fairchild, Space Industries, and Spacehab had been targeting as users for their spaceware. Fairchild had already put Leasecraft on the back burner in late 1985, for want of customers and insurance coverage.[65] Space Industries, for its part, asked NASA to stand in for the commercial companies that were disappearing. Specifically, it asked the agency to lease 70 percent of the Industrial Space Facility over five years, for a total rental of seven hundred million dollars. By so doing, the government would show itself a partner with the private sector "in the continued development of space . . . the same kind of partnership that the Government and American industry have established over the years which had resulted in American leadership and international dominance in" air transportation, computers, satellite communications, nuclear energy, and bioengineering.[66]

In fact, there were some within NASA who thought the agency could use the accommodations that Space Industries proposed to rent to them, including James T. Rose, the former head of the McDonnell Douglas electrophoresis program, who took Gillam's job at the Office of Commercial Programs in October 1987. Another vote for the facility came from materials scientist and astronaut Bonnie Dunbar, who had headed a committee on materials processing research that had issued a report in the summer of 1987. But in industry, in the Office of the Space Station, and in the office of Administrator Fletcher there was only opposition, based upon its cost to government and the fear that money spent in funding the Industrial Space Facility would threaten the very existence of the Space Station program.[67]

The firms that had received the initial eight contracts, Boeing, Martin Marietta, Rockwell, McDonnell Douglas, TRW, Rockwell subsidiary Rocketdyne, General Electric, and RCA had, in effect, bought in to the program in the hope of receiving production contracts. Many of them had spent more on the bids for the preliminary phase than the contracts had been worth. They had also signed on to small contracts to develop generic technology useful to space stations, and had put a considerable amount of their own resources into them.[68]

Industry's chance to participate in a large government aerospace program was becoming more infrequent, as federal programs became fewer, larger, and more widely spaced.[69] It is true that at the time, NASA committees and space enthusiasts were talking up a "next step" in civilian space: a permanent manned base on the moon and manned exploration of Mars.[70] This plan, however, was nowhere near to being a national commitment. The space station was and it would ensure participants the privileged access to technology they needed to remain in manned space. It would also mean contracts for billions of dollars that could extend over decades. "This is the major civilian space program for the rest of the century," Aviation Week & Space Technology quoted a Boeing vice president, "We consider it extremely important to be on board." And from McDonnell Douglas, "It's vital to us. If we don't win this, we won't be very active in manned space flight."[71]

It is true that the aerospace sector was working full tilt on military orders, but these were old orders. Détente with the Soviet Union had started the year before, and it appeared that the cold war might finally be winding down. Congress was beginning to cut defense budgets and industry was expecting a sharp contraction in the production of military planes and missiles beginning in the 1990s. The predictions were that the coming decade would see overcapacity, and that some companies would be forced into mergers or wiped out altogether.[72]

The growth areas for the industry were commercial airlines, where there was a spurt of orders,[73] and space. Sales of space hardware had been growing steadily since the mid-1970s, and in the late 1980s space hardware had become the industry's second most important product, behind military aircraft but ahead of civilian aircraft.[74] Commercial space equipment was part of the increase, although it only accounted for about 10 percent of the total U.S. space market.[75] The Strategic Defense Initiative for space-based weaponry was being funded at several billion dollars a year. Space spending as a whole was expected to continue to grow modestly, even as defense declined. Aerospace firms therefore viewed space as a hedge behind which they could take a certain amount of shelter. The NASA space station was an important part of that hedge.[76] Industry knew that space station opponents in Congress had already been advocating astronaut-tended platforms as a cheaper alternative to a manned station.[77] The Industrial Space Facility could endanger the whole station project.

Even if it did not scuttle the station, the ISF would interfere with it. The guaranteed $700 million in rent it wanted would likely come out of the station's budget. The budget for the space station, however, was already being sharply reined in. NASA had wanted $1.055 billion in fiscal year 1988; the agency had gotten $515 million instead. For fiscal year 1989, NASA had asked the administration for $1.8 billion. In December 1987, however, OMB allotted it only $967 million. By early 1988, the coveted follow-on station contracts had been awarded, to Rocketdyne, McDonnell Douglas, Boeing, and General Electric. As NASA leaders fought to get the administration request of $967 million through Congress, the last thing these firms wanted was to have some of this money diverted to rental on a free-flying space platform.[78]

But NASA was not making decisions about the Industrial Space Facility merely under pressures from industry. On the contrary, its position was more nearly akin to being on a battlefield where bullets are coming from every direction. The Department of Commerce was championing the Industrial Space Facility as an example of just the kind of private sector provision of space infrastructure that it advocated. Commerce had even succeeded in having the Industrial Space Facility and similar projects written into the new space policy that President Reagan promulgated in February 1988. Members of the House and Senate Appropriations committees of Congress also favored the ISF, some because they opposed the

station, others because they thought it important to get some sort of platform up quickly in order to counter foreign competition. On the other
hand, many members of the House and Senate Authorization committees
opposed it. After working for years to get the space station started, they
did not want to see it jeopardized by an unmanned platform. And the
aerospace companies lobbied against it because of the threat to their station contracts.[79] As part of the administration, NASA needed to show
itself bound by the president's February 1988 space policy. It therefore
proposed to run a competition for what it called a Commercially Developed Space Facility, a competition for which Space Industries would be
welcome to bid. But it structured the request for proposals to require a
facility with nearly twice the electrical power of the Industrial Space
Facility.

Most of the opposing positions were on display in an emotional set
of hearings that the House Committee on Science, Space, and Technology
held in March 1988. Max Faget and other Space Industries executives
promised they could get their ISF into orbit in 1991, four or five years before a NASA station was projected to operate. They expressed their grave
concern over the Soviet, European, and Japanese competition in the field
of microgravity research. As well as overcoming the U.S. technological
lag in that field, they explained, their facility would allow the testing of
aspects of station operation, like the use of the shuttle for resupply. The
Industrial Space Facility would be inherently more economical than a station because it was unmanned and because it could be carried up in a single launch with no need to be assembled in space. Indeed, they represented the movement from manned to unmanned system as a kind of law
of human progress: "Clearly, in fact, our human history goes away from
people doing things toward machines doing things."[80]

NASA countered that a space station had many more capabilities than
simply the microgravity research that a Commercially Developed Space
Facility could provide. It argued that an unmanned platform should, log-
ically, be a follow-up to a station and not a way station leading to it, because the automation that the platform required would not be ready until
after the processes had been run on astronaut-tended equipment. The
Commercial Space Facility would use up resources needed by the station;
money in the first instance but also shuttle flights. Finally, although NASA
"favored" a Commercially Developed Space Facility, in line with the President's policy, it favored a space station more.[81]

The Commerce Department weighed in with testimony from the Bureau of Standards, and with a message from Secretary C. William Verity
that it was vital that the president's commitment to a commercial space

facility be acted upon rapidly.[82] Committee members were divided. Robert S. Walker (R-Pa.) thought a commercial space facility "an opportunity we cannot afford to pass up," but Bill Nelson (D-Fl.) declared, "I don't care what space facility it is, in whatever form. If it gets down to a choice between that or the space station . . . this member [will choose] this Nation's necessary commitment to the space station."[83]

James Beggs also testified. The former NASA administrator, by then exonerated of the charges that had led him to step down, had taken a position as chairman of the board of Spacehab. Beggs began his testimony with a defense of his pre-*Challenger* decision in support of Space Industries' facility. As he went into the question period, however, he was pressed by David Nagle (D-Iowa), an implacable opponent of the ISF. By the end of his testimony, Beggs was representing the Spacehab as in competition for customers with the ISF. The ISF, however, would enjoy an unfair advantage because it would be propped up by NASA's guarantee to lease space from it. "The concern I have, Mr. Nagle, is simply one of providing an equal competitive opportunity . . . [if] the Government is going forward . . . with a CDSF which will be available in the same time period and could be used in a suboptimal way as competition, then I am in trouble."[84]

By summer 1988, the pro-station forces on the House and Senate authorization committees prevailed. Wording was written into the fiscal year 1989 appropriations requiring NASA to commission an outside study of a Commercially Developed Space Facility with a report due in April 1989. In May 1989, on the basis of that report, NASA canceled the CDSF competition.[85]

WHEN WE contemplate industry involvement in the station in the half decade after the *Challenger* accident, we find two opposing tendencies. Industry qua contractor got a bigger role when Level B station management hired a Grumman-led team to help with engineering and systems integration. But industry qua commercializer was forced to pull back from privately financed space infrastructure. There were two reasons and *Challenger* played a part in both. The accident made the space budget still tighter. It took money to build a new orbiter, to repair the damage to the shuttle launch pad, and to satisfy Air Force requests for expendable launch vehicles. When it looked as if the nation could afford either a station or private platforms, the big aerospace companies, and more circumspectly, NASA, fought for the station. NASA did so to retain what it saw as its core program . . . the companies to hang onto present and future contracts.

Second, the *Challenger* explosion diminished still further non-aerospace interest in renting space platforms. That, together with the rise in insurance rates that followed a string of launch failures in 1986 prompted Fairchild to drop Leasecraft. It caused Space Industries to ask NASA to take on the role of anchor tenant, making the ISF still more an anathema to station supporters. Of the four proposals here discussed, Leasecraft, ISF, Spacehab, and External Tank Company's Labitat, only Spacehab would get off the ground, in the first part of the 1990s. Spacehab made its maiden shuttle voyage in June 1993, but with 98 percent of the facility occupied by NASA experiments.[86] That is an irony that was probably not lost on the staff at Space Industries.

7

Trends in NASA-Industry Relations

NASA'S ENVIRONMENT WAS TRANSFORMED IN THE early 1990s. Among the changes were the end of the cold war, substantial reductions in the budgets of federal agencies, a massive restructuring of the U.S. aerospace industry, and novel commercial space applications. Into these altered circumstances was inserted, in 1992, a forceful new administrator, Daniel S. Goldin, who was determined to reform the agency so that it might remain viable in its changed situation.

Such a turn in NASA's affairs marks the early 1990s as the end of an epoch, and makes 1991 a natural date at which to conclude this history. At the same time, it is hard to halt completely at a turning point. We are beckoned by the road ahead, and its promise that, precisely because it has turned, it will provide us with a better perspective over the path we have been traveling. Hence this epilogue, which briefly looks at the years 1992 to 1996. In it I summarize how the momentous changes in NASA's ambience came about and how they affected the space agency. I survey some of the new policies that Goldin put into place and ask what we can say from the vantage point of recent history about trends in NASA-industry relations over the preceding three and a half decades.

We begin with an aperçu over the changes. The cold war began winding down in the late 1980s when Soviet leader Mikhail Sergeyevich Gorbachev moved to introduce greater political openness (*glasnost*) into his country and to restructure its economy (*perestroika*). The unforeseen result was a wave of restiveness that agitated the nationalities that made up the Soviet Union and the wider Soviet bloc. First, in 1989, the east European Soviet satellite nations abandoned communism. Then in 1991, the Soviet Union itself fragmented into more than a dozen nations.[1] The country that NASA had been invented to compete with in space had disappeared. To be sure, from the 1970s on, beating the Soviet Union had increasingly taken a back seat to goals such as preserving U.S. technological leadership or assuring U.S. industry a share of global markets. Nevertheless, competition with communism had remained part of NASA's raison d'être and in 1991 this rationale vanished.

The end of the cold war also brought a short-lived but intense exaltation of capitalism on a worldwide scale, a burst of enthusiasm for private as against government conduct of national affairs. It would take more research to estimate how strong the effect of this ideological surge was on the U.S. space program, but a connection between the fall of socialism and a larger role for the private sector in NASA programs was made in at least one instance, when Representative Robert S. Walker (R-Pa.), long an advocate of a greater role for the private sector in space, insisted that "it's important that we have the right attitude now, so that when the space station is built, it doesn't become the last socialist republic in the galaxy. Every republic on Earth, at least, is in the global movement toward reduced government control over economics."[2]

In the same years in which the cold war was ending, the federal budget appeared to be careening out of control. The yearly deficit—the expenditures less revenues received—had gone up sharply under the Reagan and Bush administrations. In fiscal year 1992 it hit a record $290 billion out of a federal budget of $1.38 trillion, or about 21 percent of the total. Economists in the United States and abroad voiced alarm. So did the U.S. citizenry, which became still more exercised over the issue in 1992 when Ross Perot, a Texas businessman who had never held an elected political office, made ·it the centerpiece of an independent campaign for the presidency. Congress and the administration began cutting domestic programs to bring down the deficit and the mounting national debt that accompanied it. NASA programs were not exempted. NASA's budgets had increased, in inflation-adjusted dollars, during the 1980s. In the 1990s that trend was reversed as they first stabilized, and then began to decline.[3]

The aerospace industry was hard hit by the budget cuts. It was not simply NASA, but more importantly, the Department of Defense, whose budget was twenty-fold larger than NASA's. Added to this was an unexpected downturn in the market for commercial airliners in the early 1990s, affecting principally Boeing and McDonnell Douglas.[4]

The commercial airline business was in the grip of a structural transition, moreover, as well as a temporary slump in sales. The market dominance of U.S. firms was being broken. A European consortium, Airbus, had slipped into second place in terms of market share, behind Boeing but ahead of McDonnell Douglas. Pacific rim nations like Taiwan, South Korea, Singapore, Japan, and Indonesia were moving to develop indigenous airplane industries. The prospect for the future was that the global market for airlines would no longer be divided among U.S. firms, but among firms from several nations.[5]

Faced with these conditions and pressed by their stockholders to maintain profits, some of the large U.S. companies left the aerospace industry. GE sold its aerospace division to Martin Marietta. Longtime NASA contractor Rockwell International sold its aerospace and defense units to Boeing and concentrated on electronics and computers. General Dynamics' missiles business went to Hughes. Other companies merged. Northrop took over Grumman. Lockheed and Martin Marietta combined to form Lockheed Martin. Boeing began the process of absorbing McDonnell Douglas.

The consequence for NASA was a much diminished group of firms to which it could turn when its projects called for a large prime contractor with wide-ranging capabilities. NASA faced the prospect that for such contracts, it might receive only two, or even one bid. (This happened in 1997 when Boeing temporarily declined to bid on a six billion dollar contract, leaving Lockheed Martin the sole bidder.)[6]

At the same time, the outlook for space was not as bleak as the outlook for other aerospace products. Defense spending for space was declining less than for missiles or aircraft. Total space spending—military, civilian, and commercial—was holding steady.[7] The reason was that technological advances were opening new uses for space systems.

These uses were both military and commercial. On the military side, research by the Defense Advanced Research Projects Agency and other Defense Department bureaus into small satellites had begun to bear fruit, and it was becoming possible to conceive of adding fleets of small satellites dedicated to a single commander in a single battlefield to the large, multipurpose satellites whose use was shared by U.S. commanders over the entire globe. Such usage would be practical if satellites became truly cheap—enough so that commanders could train their troops with satellites beforehand—and if launching procedures became rapid enough that large numbers of satellites could be fielded as soon as a crisis erupted.[8]

In the forefront of new commercial applications were novel uses for surveillance satellites. Here the technological advances included better sensors, space platforms, and telemetry systems and an improved understanding of how the earth's systems interact with sunlight. This made it possible to use satellites to supplement aerial and balloon photography to do things like chart vegetation, monitor the earth's crustal movements, locate schools of fish or fleets of trucks, collect data for maps, and prospect for oil and minerals. Remote sensing from satellites thus held promise of being useful to industries like cartography, agriculture, mining, fishing, and transport.

New opportunities were also emerging for communications satellites. The growth of the Internet meant new communications links were in demand, and satellites were especially useful for linking computers in areas where laying fiber optics cables was uneconomical. As antennas became small enough for use in urban as well as rural areas, there were expectations that the market for direct broadcasting of TV programs from satellites to homes would grow. Communication via satellite with mobile users, in boats, planes, and trucks, was becoming increasingly important.

Established aerospace firms were moving into some of these fields. Lockheed Martin, Ball Aerospace, Orbital Sciences, and Boeing all joined remote sensing ventures.[9] In addition, small firms were started. In remote sensing alone, the number of U.S. companies involved nearly tripled between the beginning and end of the 1980s.[10] Even the one area in which technological progress was lagging—finding ways to get payloads into orbit more cheaply than the thousands of dollars per pound it still cost—was stimulating entrepreneurs. New companies were forming around innovative ideas for space transportation. Pioneer Rocketplane, Kistler Aerospace, and Kelly Space and Technology, Inc., were just some of the companies the press was beginning to cover.[11] Thus, and somewhat paradoxically, at the very moment at which NASA had fewer large U.S. aerospace companies for big contracts, it had plenty of contractors to choose among for small-scale projects.

GOLDIN CAME into this picture in early 1992, having been appointed after a long series of policy disputes between NASA Administrator Richard Truly and Vice President Dan Quayle and his National Space Council. In 1989 Truly and the administration crossed swords over the Space Exploration Initiative. In addition, Truly had little enthusiasm for two new launch technologies, the National Aerospace Plane, a joint project of NASA and the Defense Department, and the National Launch System, an expendable launcher Quayle had promoted. Quayle also wanted NASA to incorporate some of the new technologies and management practices that were being tried in industry and the military. He was pushing NASA to replace its large science projects with smaller, cheaper spacecraft that could be designed and built more quickly. Ex-astronaut Truly, on the other hand, focused on the shuttle and the space station, NASA's core human flight programs. He did begin putting in place one of the new management techniques then in vogue, total quality management, but did not share Quayle's enthusiasm for reaching outside NASA for innovative ideas. In February 1992, Bush asked Truly to resign. Goldin, the man brought in to replace him, was a TRW executive who had mainly managed the company's highly secret, unmanned satellite projects for intelligence agencies.[12]

Quayle left office with Bush after the 1992 elections, but the new president, William J. Clinton, kept Goldin on. Complementing him, in the White House, was John Gibbons, the savvy ex-head of the Congressional Office of Technology Assessment, whom Clinton appointed to lead a new Office of Science and Technology Policy. The more extravagant of Quayle's plans had already been shaved away. The Moon-Mars initiative had been stillborn, with Congress from the first balking at the price tag NASA had put to it.[13] The National Aerospace Plane and the National Launch System were moribund. But Goldin retained Quayle's thrust toward innovation.

As NASA administrator Goldin faced all the problems we have noted. The cold war justification for NASA's programs had vanished. Congress and the administration were signaling that NASA's budget, which had risen from six to fourteen billion dollars over the prior decade (in 1994 dollars, from nine billion to fifteen billion dollars), would level off or decrease. The U.S. aircraft industry was losing market share to foreign companies. In addition, after a decade of work and the expenditure of billions, the space station program had produced little but paper. The shuttle was functioning well, but had not solved the problem of frequent and cheap transportation into space. Shuttles were flying about seven times a year, as opposed to the weekly or biweekly flights predicted in the early 1970s, and launch costs were a half billion dollars per flight, as against the seventy to one hundred million dollars that had been projected. Competition in the launch industry was increasing as Russia, China, and Japan started to market launchers in contest with U.S. launchers and Ariane, while U.S.-made launchers were based on old technologies and were overly expensive.

Goldin's solution was "nonlinear thinking." NASA was going to spend less, but it was not going to do less. Instead, it was going to do things differently. This included smaller satellites sent up by cheaper rockets and more use of new technology. Most relevantly for our themes, it included privatization of NASA activities wherever possible, smaller management teams, and cost-sharing with industry.[14]

The station was the first order of business for Goldin. In 1993, he selected a single prime contractor, Boeing, to oversee the project, and eliminated the NASA Level B management. The prior system of three or four work packages, each managed by a different NASA center and each with its own "prime" contractor, had been severely criticized by Congress: too much of the station budget was being consumed in managing interactions among the centers. It was an arrangement in which NASA was not so much acting as its own prime as housing within itself a number of competing primes.[15]

The use of a single prime contractor had been recommended by the Augustine Commission. It was the old Air Force "weapons system" practice, discussed in chapter 2, which gave an aerospace firm wide cognizance over systems integration and the work of subcontractors. As a first result NASA personnel supervising the station dropped from 3,000 to 1,300. The change also opened the way to the introduction into NASA of some of the new management practices Quayle had wanted, for Boeing was in the forefront of these practices. The company brought with it "integrated product teams," a part of total quality management in which design engineers, operations personnel, and manufacturing engineers work together on all aspects of a program, from design onward. Boeing also applied commercial practices to tighten financial control and generate more realistic estimates of costs.[16] Shortly after, in September 1993, the United States brought in Russia as an additional foreign partner on the space station. From the administration's perspective, the move was useful in forging a cooperative modus vivendi with the new Russian state. From NASA's point of view it was a twist to an old theme—the Russian Bear was once again pressed into service to help sustain a NASA program.

Changes in shuttle management came next, triggered by the tightening budgets. NASA had spent $4.3 billion in fiscal year 1991 on the shuttle, nearly a third of its budget. Part of this had gone into upgrades to make the vehicle capable of carrying construction materials for the station. Much, however, was for operations, and Goldin thought operations consumed money that was better spent for new, innovative programs. At the beginning of 1995, President Clinton ordered NASA, which had been expecting level budgets to the year 2000, to cut spending by five billion dollars over the half-decade. At this point the cost of shuttle operations became untenable. An outside review panel recommended, and NASA accepted, a change from NASA management of shuttle operations to management by an industrial contractor. The NASA team supervising the shuttle could then be decreased from where it stood at nearly 3,000. Lockheed Martin and Rockwell put together a joint venture, under the name United Space Alliance, which took over in October 1996. United Space Alliance spokesmen portrayed their company's role under NASA supervision as similar to an airline's operation of commercial aircraft under the oversight of the Federal Aviation Administration.[17]

The problem of low cost, routine transportation into space was judged on all sides to be of overriding importance. A White House National Space and Technology Council wrestled with the issue in 1994; Goldin and Gibbons took leading roles. The presidential National Space Transportation Policy that resulted from the council's work divided the job of developing better space transportation between the Department of Defense and NASA. The DoD was charged with creating a family of unmanned "evolved expendable launch vehicles" (EELV) that would be cheaper to fly. NASA took on a program that was far more innovative technically, managerially, and financially, that had as its goal a reusable vehicle that could replace the shuttle early in the twenty-first century.[18]

Technically, NASA chose a completely reusable single-stage vehicle, designed for "dual use," that is, to satisfy at once the needs of government and the commercial launch market. Financially, NASA and the companies entering the program were both to provide funding. Managerially, the program was to be carried out by industry-led "synergy teams" composed of industry, NASA, and Air Force personnel. To keep development costs low, NASA took a leaf from Department of Defense's book and set up a headquarters staff of only twelve people, and it brought in a DoD manager, Colonel Gary Payton, to head the overall effort.[19]

Joint financing was appropriate, because the project was directed toward a vehicle that private industry would own and, if all went well, would profit from. Joint financing was also indispensable because NASA simply did not have the money to fund the reusable launch vehicle (RLV) program by itself. "Partnership" was very much on the lips of NASA executives. "We are partners with industry and not their managers," Associate Administrator for Space Access and Technology John Mansfield told the Senate Committee on Commerce, Space, and Transportation.[20] But it was partnership with a whole new meaning. Rather than being a term that justified industry's obtaining financial favors from the government, as in the 1980s, it now lent legitimacy to NASA's seeking funding help from industry.

The NASA RLV program comprised several distinct projects. From the Ballistic Missile Defense Organization, the successor to the Strategic Defense Initiative Organization, NASA took over the DC-X, a single-stage-to-orbit rocket that McDonnell Douglas had been building. Under the new arrangements, NASA would put up 70 percent of the funding of a larger DC-XA, and McDonnell Douglas 30 percent. A second component of the RLV program was the X-34 project, for a small, partly reusable, rocket that could serve NASA as a test bed for new technologies and at the same time serve the companies that made it as a commercial launcher. NASA was to put seventy million dollars into this project and the companies one hundred million dollars. In March 1995, NASA signed an agreement with Orbital Sciences Corporation and Rockwell International to comanage the X-34.[21]

The centerpiece of the program was the X-33, a completely reusable, single stage vehicle, but not an operational one. It would be subscale, pilotless, and would not be required to go into orbit. Its function would be to prove the feasibility of building an inexpensive single-stage vehicle and show that it could operate with small launching crews, have fast turn-around, and be easy to maintain.

The X-33 program was divided into three phases. Phase I, begun in March 1995, awarded contracts to three industry teams, headed by Lockheed Martin, McDonnell Douglas and Boeing, and Rockwell, for paper designs. NASA was paying for the design process but—and this was an important novelty—it was not dictating the design. Instead the teams were instructed to propose their own concepts. It was a return to the old practice of securing hardware through a shoot-out among competing industry designs. Lockheed Martin elaborated a wingless delta-body plane whose shape was reminiscent of the Lockheed STAR Clipper of the late 1960s. The team of McDonnell Douglas and Boeing chose to develop a version of the DC-X. Rockwell worked up a design derived from the shuttle.

In July 1996, NASA selected the Lockheed Martin design, called VentureStar, and Phase II started. NASA is slated to put up about $900 million for this part of the program, with Lockheed Martin contributing a minimum of $200 million.

The X-34 program, meanwhile, had shipwrecked as a jointly financed project. After a rough voyage through 1995 and early 1996, marked by controversies among the government and company managers, Orbital Sciences pulled out in February 1996. The costs to develop the vehicle were soaring and Orbital Sciences was unable to get its NASA and Rockwell partners to agree to pursue a cheaper design. NASA then reformulated the program, deleting the requirement that the vehicle be commercially useful and leaving it simply as a technology test-bed. The requirement for a financial contribution from the contractor was also removed. Orbital Sciences Corporation, Rockwell, and McDonnell Douglas competed for this new government-financed project and Orbital Sciences won.[22]

As regards the X-33, aerospace companies have been eager to enter Phase II, where the government is to put up $900 million, but they have expressed doubt about Phase III. Phase III calls for a full scale RLV to be built, financed, and operated entirely by private enterprise. On this scheme, then, the successors to the shuttles would be "privatized" from the start. They would be commercial spaceliners and space transports on which the government and everyone else could book their rides and their cargoes. Industry and outside critics have questioned the fundamental premise of dual use, namely whether the craft can really be engineered to satisfy both NASA's need for a ship to service the space station and the commercial need to loft satellites. Launch vehicles for satellites can be made much cheaper if they do not have to be safe enough for human flight. And if the industry balks at the end, will the taxpayers wind up paying for the shuttle's successor after all?[23]

Industry has also pointed out that the U.S. market for launch vehicles is largely dominated by the Department of Defense and NASA: when the new reusable vehicle finally comes on line, the government is likely to still be the major customer. In an amusing rerun of the early 1970s, when NASA fought to get the Defense Department to commit all its payloads to the shuttle, industry has asked both DoD and NASA to place all their cargoes on any RLV they might finance. "During the critical early years of operation . . . we must be assured the U.S. government will commit its market to us if we provide a superior product," testified one industry executive. And a NASA official echoed the thought, "What would devastate them, is if they had the ability to launch U.S. government satellites and, for reasons of national security, one thing or another, we kept other lines open. They need a guarantee that . . . we will not find another reason to put the business elsewhere."[24]

IF WE look back at NASA-industry relations from the vantage point of recent years, what is most striking is how slowly things changed during the first three decades, in comparison with the Goldin years. This is true whether we look at the design of manned space craft, the practice of cost-sharing, dual technology, or other areas. In the design of manned craft, industry gradually took a greater role from the 1950s through the 1980s. The design for the Mercury capsule was done in-house in a period of months by a small team under Maxime A. Faget. With the Apollo capsule NASA decided to commission preliminary design studies as part of a move to enhance industry capabilities and three companies were given contracts for them. But in the end, NASA simply chose to go with its own Apollo design—once again put together by a team under Faget.

This pattern was broken with the shuttle. There was a Faget design, but this time it was peripheral. NASA no longer had the luxury of taking only its own ideas into consideration. There was a protracted fight to secure funding for the shuttle and the design process slowed until it stretched over two years. In these circumstances, industry, which by now had a decade of NASA-and-military contracts on shuttle-like vehicles behind it, had time and opportunity to feed its own suggestions to the agency. For the space station these circumstances repeated themselves in spades. Now industry had two decades of prior study contracts under its belt and a ten-year span in which to insinuate its ideas into NASA planning.

There is, however, a difference in kind between even a measurably increased flow of industry ideas into NASA designs and a situation like the X-33, where industry elaborated the designs and NASA chose among them. In the latter case, NASA ceased to act as its own design house and took up a role more like that of the military. The X-33 is not the only case of this practice in the new NASA. Another recent example is the Bantam project, where NASA is funding four industry teams to pursue small-launcher designs that originated inside the companies.[25]

Cost-sharing with industry and technologies designed for dual government and commercial use were not untried in the old NASA. Witness the ACTS program, where, as we have seen, NASA originally proposed that the satellite that would carry the ACTS instrument package be used simultaneously for experiments of the contractor's choosing. The aim of this proposal, in the event one that was not achieved, was to get the contractor to share the cost of the satellite with NASA. Witness also the attempts to design a space station that would be suitable for military, commercial, and scientific purposes.

These were mere adumbrations, however, compared to the use of cost-sharing and dual technology in Goldin's NASA. The X-34 and X-33 are outstanding examples, but there are other instances. LightSAR is a project that, like ACTS, is intended to support commercial enterprise. It is an experiment in remote sensing using radar, as opposed to optical or infrared frequencies. NASA has attempted to get industry to finance up to 50 percent of LightSAR. In what might be thought of as a reversal of dual technology, NASA has also tried to get producers of commercial satellites to modify their spacecraft so that NASA might use them for its own remote sensing programs.[26]

The immediate spur to cost-sharing and dual technology alike is, of course, NASA's tight budget. But both practices also have to be seen in the context of a broad interest within the Clinton administration in dual technologies, and a general movement in business to cost-sharing. NASA is not immune to the influence of changing business practices. On the contrary, the winds that bring new management techniques and management fads sweep equally across industry and government; total quality management is an example. Cost-sharing is both fad and technique. It is a fad in that U.S. business writers and politicians alike, in the 1980s, were smitten with the model provided by the government-industry consortia established in Europe and Japan. It is technique in that today, joint ventures with shared financial risk are a way of life in national and international business. NASA has, of course, structured its International Space Station as a joint venture of governments, but it has also used the joint venture as a template for industry-NASA relations.

One usage where little changed through 1991 was NASA's insistence on looking over all details, its "technical penetration" into contractors. Although industry came to do more of the work, taking over, for example, more and more of the testing,[27] it was still saddled with small armies of NASA managers. The Augustine Committee spoke to that in its 1990 report when it called attention to the Department of Defense's space program, almost twice the size of NASA's, but operated with only limited in-house personnel.[28]

The new NASA has been trying to alter this practice. The small headquarters staff overseeing the reusable launch vehicle program has already been mentioned. Another instance is the Lunar Prospector, a "smaller, cheaper" satellite (budgeted at sixty-three million dollars), that will map the moon. This program is being run out of the Ames Research Center. "The only way you can do a very cost-constrained program," the NASA manager has been quoted as saying, "is by letting the contractor focus on doing the job, taking responsibility for it, and not having to respond to a great deal of shadow organization from the NASA entity."[29]

A similar situation obtains in privatization. There was a good deal of talk about privatizing the shuttle before 1991 and there was talk of incorporating privately owned elements into space infrastructure like the shuttle or station. But little happened. The Boeing logistics modules were never incorporated into the station. The Industrial Space Facility was perceived by NASA and the aerospace industry to pose a threat to the station program, and was defeated. Klaus Heiss's and Willard Rockwell's plans for privately funded shuttles were judged to be premature and unsound even by those at NASA most sympathetic to privatization. Only Space-hab's project, which placed privately owned experimental modules in the shuttle, was still alive in the planning stage by the end of 1990.

In contrast, under Goldin, shuttle operation has been turned over to a contractor. More private companies are being let into the processing of data from the environmental monitoring satellite system that NASA is preparing. A mammoth Consolidated Space Operations Contract is being structured that would turn over to a private firm a raft of NASA operations, even including the operation of mission control centers for human flight.[30] And these are only some of the privatization moves underway or contemplated.

This chapter has not sought to examine the industrial policy of Goldin's NASA—its approach to fostering the commercial space industry. Even without the vantage point that knowledge would bring, however, NASA's industrial policy through 1991 appears singularly ad hoc. NASA aided the communications satellite industry from the start and, in the early years, exercised considerable influence over the shape of the technology and the ability of private companies to succeed. On the other hand, NASA resisted the buildup of a commercial launch industry. Launching was for many years an enterprise that was run by a de facto partnership of NASA and the companies from which NASA bought launchers and launch services. NASA proposed to put an end to that enterprise in the 1980s; it sought to enthrone the shuttle as the nation's commercial, as well as government, launcher. The prospect of erecting a private sector launch industry along-side the NASA shuttle was discussed, but it did not become a reality because the shuttle was too tough a competitor for private vehicles. Only the grounding of the shuttle after the *Challenger* accident allowed the commercial launch industry to get started.

NASA promoted space industrialization, particularly materials processing in space (MPS), from the 1970s on because it promised cargoes for the shuttle and support for the space station. The aerospace industry also tried to recruit nonaerospace firms for space-based manufacturing and attempted to mount a few projects of its own. Aerospace had the same goals as NASA: to maximize shuttle use and to ensure that the station was funded. Beyond that, it saw a market in building equipment and space platforms for space ventures. The new space shuttle was expected to lower drastically the cost of putting MPS experiments and manufacturing facilities into orbit. By 1990, however, it was clear space manufacturing had been oversold. The shuttle was not providing the cheap access to space that had been promised, and that was a sine qua non for cost-effective space manufacturing. No product had been identified that was valuable enough to command the high price companies would have to charge. Earth-based processes were popping up to compete with space processes for manufacture of some of the more promising products. Even MPS enthusiasts now foresaw a long period of experimentation, with commercial uses, if any, far in the future.

These cases all indicate a "policy" that was never thought out as a whole, and that was too much influenced by the needs of NASA's own programs or groups. All three policies, it must be pointed out, date from before 1980. After 1980, a conscious examination of commercial policy began, first as part of the process that led to the Office of Commercial Programs, and after 1985, both within that office and in the administration of Reagan and Bush. It would be useful to study the evolution of this thinking, the influences on it, and its effect on NASA. At a first glance, however, it appears to have had no major consequences through 1991.

AN UNEXAMINED assumption often figures in the writings of members of the space community or of the outsiders who study it: that the passage from government to private dominance of space is natural or even inevitable. Sometimes this assumption takes the form of a reference to Christopher Columbus and Queen Isabella and the observation that early exploration of new frontiers have traditionally been funded by governments, while exploitation of these frontiers has been reserved for later-coming private entrepreneurs. Sometimes it is tied to examples from U.S. history, the development of the railroad system, where the federal government gave industry subsidies and huge grants of public lands; or the airplane, where commercial airliners were often based upon military planes developed at government expense. Sometimes the appeal is to peculiarly American ways of doing things, an indigenous culture of government-industry interaction that gives as much scope as possible to the private sector.[31]

This way of viewing things often goes along with the idea that the progression from public to private has been delayed or resisted by NASA's organizational bent. NASA is seen as peopled by engineers whose hearts are in research and development, rather than in commercial aspects of space. NASA managers are pictured as inheriting a tradition from the Apollo era that makes cost no object, whereas cost is paramount in industry. NASA is supposed to be adverse to risk, a posture that reflects, at least in part, its role as public servant, publicly accountable for the sound use of taxpayer dollars. Private capitalist industry is risk-taking by nature.[32]

In point of fact, we *have* been witnessing a transfer of space activities from the government to the private sector, gradual over the period from

1958 to 1991, more rapid in the last five years. But what has been "natural" or unavoidable about this transformation and what has been fortuitous? One place where proponents of the natural view have some ground is in the diffusion of technical expertise to more and more industrial firms. Metaphor is at work here, of course. The idea of an ever-widening diffusion of space technology conjures up the second law of thermodynamics, with its fearful inevitability. But reality in this case appears to match well with image. The spread of knowledge, both within a nation to more and more organizations, and internationally, may be as close to unavoidable as social phenomena get. Certainly, as we saw in chapter 2, it was a deliberate strategy of T. Keith Glennan and his McKinsey and Company advisors to educate industry so as to foster this transfer of activities. (We may wish to remember, however, that the model of knowledge as flowing unidirectionally from NASA to industry is probably inadequate. NASA was no less a novice in matters of space in 1958 than industry was, and the companies both taught the agency and learned from it.)

Another phenomenon with some claim to the classification "natural" is the parade of new commercial applications for space. Again, there is a metaphor at work, this time from the organic world. A new technology seems like an embryo, with its prodigious capacity to develop. The fact that there are increasing commercial opportunities in communications, remote sensing, and navigation is certainly one of the causes for the increasing involvement of private firms in space activities.

So these are two good reasons to embrace the idea of a natural progress from space as a government activity to space as a private activity. Even here, however, we need to be alert to the influence of special pleading or special points of view. Industry has often pressed for a greater role in the U.S. civilian space program. It gives companies more control, more contract money, and more opportunity to build in-house technical capability. When, therefore, advisory committees composed largely of aerospace executives argue that there are natural reasons that contractors should receive more of the tasks in space projects, it is worth recalling that this way of viewing the matter may be congenial to their interests or to their underlying worldview.

If a diffusion of expertise and a multiplying number of commercial possibilities as factors pushing for a transfer of space activities to the private sector seem unavoidable, however, there are plenty of causes for that transfer that appear strictly contingent. One of these is the shift in political ideology, from the Kennedy-Johnson years when the administration and polity supported an activist government, to the Reagan administration, with its antigovernment rhetoric and exaltation of private industry.[33] "Small government" ideology helped swell enthusiasm for commercial space ventures in the early 1980s. It was a factor in the history of NASA's ACTS program and it led to increased discussion of and proposals for privatization of NASA's space infrastructure. There was nothing inevitable in this ideological shift and there is no reason why it might not reverse itself.

A second fortuitous factor was NASA's movement from a central instrument of administration policy during the Kennedy years to its status as a peripheral agency under Nixon. Having lost its role as a weapon in the cold war, and in a context in which economic conditions were worsening, NASA was forced to justify its programs on the grounds of their contribution to the nation's economic well-being and competitiveness. In these circumstances, NASA could do no other than manifest increasing concern for commercial space in its pronouncements and its activities.

The circumstance that the Reagan and Bush administrations ran up huge budget deficits can hardly be classed as unavoidable. Yet the results this engendered have had a pronounced impact on NASA's relations with industry. In its general downsizing of government, Congress has not particularly wanted to slash NASA funding.[34] Even the most budget-minded of Republican members have often had a warm spot for NASA and its contractors. But they have seen no alternative. Budget reductions have been the proximate cause for turning the shuttle over to United Space Alliance, and a major reason for seeking industry financing for the reusable launch vehicle program.

Finally, we have seen that much of the NASA-industry interaction was simply shaped by the interests—sometimes parochial and sometimes altruistic—of groups within the agency or individual firms. Into this category go the opposition of aerospace firms to the Industrial Space Facility, seen as a possible threat to space station contracts, NASA's own antagonism to expendable launch vehicles, and Hughes' attempt to shoot down the ACTS program. One can hardly pin labels like "natural" or "inevitable" to such causes.

Dealing with questions like what is necessary and what is accidental in NASA-industry relations may be one contribution historians can make to a space policy debate. In a recent, insightful book, political scientist Roger Handberg gives one analysis of this type.[35] Handberg argues that the cold war frustrated private efforts to commercialize space in a number of ways. First, national security concerns, nationalist buying practices, and the bipolar division of the world hindered business' ability to form alliances with foreign companies and to sell abroad. Second, private sector research was skewed toward satisfying the technical agenda of the public sector, an agenda not particularly congruent with commercial needs. Third, NASA was given the chief responsibility for commercialization policy even though NASA's culture was inimical to commercialization. Finally, companies became slothful, dependent on government subsidies, and satisfied to follow whatever direction government funding marked out.

Once the cold war ended, Handberg continues, companies were deprived of Defense and NASA contracts and forced to take commercial space more seriously. At the same time, they became freer to make the cross-national deals that have become essential to late twentieth-century business. The result is that the private space sector is, and necessarily will be, escaping more and more from the dominance of government. The main explanatory agent in Handberg's account is therefore the cold war, a contingent development. Its rise caused the government to take the leading role in space and its termination is moving the private sector to center stage.

Whether one agrees or disagrees with Handberg's analysis, one must concede the importance of his attempt for public discourse. Of course, as Handberg points out and other commentators have also noted, there *is* no public debate on space policy. Instead, and here NASA and industry public relations bear some blame, the public treats space projects as so many TV spectaculars, to be enjoyed rather than discussed. Still, a public debate may yet happen, and when it does, it will be sounder to the degree that historians and other scholars have worked through these issues.

Notes

Chapter 1 Partners in Space

1. This sets aside the vital NASA missions of science and aid to the U.S. military, as well as the abiding mission of space buffs inside and outside NASA of human exploration of the solar system because "it's in our genes." On the latter see Roger D. Launius, "Early U.S. Civil Space Policy, NASA, and the Aspiration of Space Exploration," in *Organizing for the Use of Space: Historical Perspectives on a Persistent Issue*, Roger D. Launius, ed., AAS History Series, Vol. 18 (San Diego: Univelt, 1995), 63–86.

2. For a study of the purposes of Apollo, see Vernon Van Dyke, *Pride and Power: The Rationale of the Space Program* (Urbana: University of Illinois Press, 1964).

3. See, for example, JoAnne Yates, "Co-evolution of Information Processing Technology and Use: Interaction between the Life Insurance and Tabulating Industries," *Business History Review* 67, no. 1(Spring 1993):1–51; David C. Mowery, *Alliance Politics and Economics: Multinational Joint Ventures in Commercial Aircraft* (Cambridge, Mass.: Ballinger, 1987); and Ronald Kline and Trevor Pinch, "Users as Agents of Technological Change: The Social Construction of the Automobile in the Rural United States," *Technology and Culture* 37(1996):763–95.

4. For the government vis-à-vis the pharmaceutical industry, see Louis Galambos with Jane Eliot Sewell, *Networks of Innovation: Vaccine Development at Merck, Sharp & Dohme and Mulford, 1895–1995* (Cambridge: Cambridge University Press, 1995). For the military and maser and laser companies, see Paul Forman, "Atomichron[r]: The Atomic Clock from Concept to Commercial Product," *Proceedings IEEE* 73(1985):1181–1204. For two economists' expression of

the importance of networks, see Richard R. Nelson and Nathan Rosenberg, "Technical Innovation and National Systems," in *National Innovation Systems: A Comparative Analysis*, Richard R. Nelson, ed. (New York: Oxford University Press, 1993), 15.

5. Leonard S. Reich, *The Making of American Industrial Research: Science and Business at GE and Bell, 1876–1926* (Cambridge: Cambridge University Press, 1985); Edward W. Constant, II, "The Social Locus of Technological Practice: Community, Systems, or Organization?," in *The Social Construction of Technological Systems: New Directions in the Sociology and History of Technology*, Wiebe E. Bijker, Thomas P. Hughes, and Trevor J. Pinch, eds. (Cambridge, Mass.: MIT Press, 1987), 223–42.

6. Richard H. K. Vietor, *Contrived Competition: Regulation and Deregulation in America* (Boston: Belknap Press of Harvard University Press, 1994); William A. Niskanen, *Bureaucracy: Servant or Master? Lessons from America* (London: Institute of Economic Affairs, 1973).

7. This definition is taken from Ha-Joon Chang, *The Political Economy of Industrial Policy* (New York: St. Martin's, 1994).

8. Stephen Wilks and Maurice Wright, eds., *Comparative Government-Industry Relations: Western Europe, the United States, and Japan* (Oxford: Clarendon Press, 1987).

9. David DeVorkin, *Science With a Vengeance: The Military Origins of Space Sciences in the American V-2 Era* (New York: Springer-Verlag and Oxford University Press, 1992).

10. See J. S. Dupré and W. E. Gustafson, "Contracting for Defense: Private Firms and the Public Interest," *Political Science Quarterly* 77(1962):161–77. On the defense industry, see William L. Baldwin, *The Structure of the Defense Market, 1955–1964* (Durham, N.C.: Duke University Press, 1967); Herman O. Stekler, *The Structure and Performance of the Aerospace Industry* (Berkeley: University of California Press, 1965); and Jacques Gansler, *The Defense Industry* (Cambridge, Mass.: MIT Press, 1982).

11. Thomas J. Misa, "Military Needs, Commercial Realities, and the Development of the Transistor, 1948–1958," in *Military Enterprise and Technological Change: Perspectives on the American Experience*, Merritt Roe Smith, ed. (Cambridge, Mass.: MIT Press, 1985), 253–87; Ernest Braun and Stuart MacDonald, *Revolution in Miniature: The History and Impact of Semiconductor Electronics* (Cambridge: Cambridge University Press, 1978); Joan Lisa Bromberg, *The Laser in America, 1950–1970* (Cambridge, Mass.: MIT Press, 1991).

12. Brian Balogh, *Chain Reaction: Expert Debate and Public Participation in American Commercial Nuclear Power, 1945–1975* (Cambridge: Cambridge University Press, 1991), chap. 4.

13. Tom Kemp, *The Climax of Capitalism: The US Economy in the Twentieth Century* (London: Longman, 1990); Philip Armstrong, Andrew Glyn, and John Harrison, *Capitalism Since 1945* (Oxford: Basil Blackwell, 1991).

14. Chang, *Industrial Policy*; Armstrong, Glyn, and Harrison, *Capitalism since 1945*.

15. Robert M. Collins, "Growth Liberalism in the Sixties: Great Societies at

Home and Grand Designs Abroad," in *The Sixties . . . from Memory to History*, David Farber, ed. (Chapel Hill: University of North Carolina Press, 1994), 11–44.

16. Dwayne A. Day, "Invitation to Struggle: The History of Civilian-Military Relations in Space," in *Exploring the Unknown*, Vol. 2, *External Relations*, John M. Logsdon, ed. (Washington, D.C.: NASA, 1996), 233–56.

17. The example is the McDonnell Company's F-4 Phantom II, as explicated in Glenn E. Bugos, "Manufacturing Certainty: Testing and Program Management for the F-4 Phantom II," *Social Studies of Science* 23(1993):265–300.

18. Senator Adlai E. Stevenson, "The New Era in Space," *Journal of Contemporary Business* 7, no. 3(1978):7–12, quotation on 7.

19. NASA Administrator James M. Beggs, testimony, House Committee on Science and Technology, "Space Commercialization," 98th Cong., 1st sess., May 1983, 26.

20. David H. Langstaff, chief financial officer, Space Industries, Inc., testimony, House Committee on Science, Space, and Technology, "1989 NASA Authorization," 100th Cong., 2d sess., March 1988, 479.

21. NASA Acting Associate Administrator for Advanced Concepts and Technology, Gregory M. Reck, testimony, House Committee on Science, Space, and Technology, "NASA's Commercial Space Programs," 103d Cong., 1st sess., October 20, 1993, 8.

22. White House Office of Science and Technology Policy Director John H. Gibbons, testimony, House Committee on Science, Space, and Technology, "National Space Transportation Policy," 103d Cong., 2d sess., September 20, 1994, 37; and NASA Associate Administrator for Space Access and Technology John Mansfield, testimony, Senate Committee on Commerce, Science, and Transportation, "The NASA Space Shuttle and the Reusable Launch Vehicle Programs," 104th Cong., 1st sess., May 16, 1995, 12.

Chapter 2 Legacies

1. This is the phenomenon of "technological convergence" described in David C. Mowery and Nathan Rosenberg, *Technology and the Pursuit of Economic Growth* (Cambridge: Cambridge University Press, 1989), pt. 4.

2. "National Aeronautics and Space Act of 1958" in *Exploring the Unknown: Selected Documents in the History of the U.S. Civil Space Program*, Vol. 1, *Organizing for Exploration*, John M. Logsdon, ed. (Washington, D.C.: NASA, 1995), 334–45.

3. John E. Burchard, ed., *Rockets, Guns and Targets: Rockets, Target Information, Erosion Information, and Hypervelocity Guns Developed During World War II by the Office of Scientific Research and Development* (Boston: Little, Brown, 1948); Roger E. Bilstein, *The American Aerospace Industry: From Workshop to Global Enterprise* (New York: Twayne, 1996), 109–17. Andrew G. Haley, *Rocketry and Space Exploration* (Princeton, N.J.: Van Nostrand, 1958). I am indebted to J. D. Hunley for this reference.

4. "Rockets," *Fortune* 42(November 1950):118–20. J. D. Hunley, "The Evolution of Large Solid-Propellant Rocketry in the United States," *Quest: The History of Spaceflight Quarterly* 6, no. 1(1998):22–38. David Dyer and David B.

Sicilia, *Labors of a Modern Hercules: The Evolution of a Chemical Company* (Boston: Harvard Business School, 1990).

5. Clayton R. Koppes, *JPL and the American Space Program* (New Haven: Yale University Press, 1982), chap. 2. David H. DeVorkin, *Science with a Vengeance: The Military Origins of the U.S. Space Sciences in the American V-2 Era* (New York: Springer, 1992), chap. 4.

6. Ernst Stuhlinger and Frederick I. Ordway, III, *Wernher von Braun, Crusader for Space: A Biographical Memoir* (Malabar, Fla.: Krieger, 1994).

7. G. R. Simonson, ed., *The History of the American Aircraft Industry* (Cambridge, Mass.: MIT Press, 1968), 181. John B. Rae, *Climb to Greatness: The American Aircraft Industry, 1920–1960* (Cambridge, Mass.: MIT Press, 1968), 81.

8. Rae, *Climb to Greatness*, chap. 9.

9. On nuclear propulsion, see R. W. Bussard and R. D. DeLauer, *Fundamentals of Nuclear Flight* (New York: McGraw-Hill, 1965). W. Henry Lambright, *Shooting Down the Nuclear Plane*, Inter-University Case Program #104 (Indianapolis: Bobbs-Merrill, 1967).

10. J. Leland Atwood, interview by Martin J. Collins, transcript, January 19 and 20, 1989, in Glennan-Webb-Seamans-Project Interviews, National Air and Space Museum, Washington, D.C., 16–34. North American's moves occurred against the background of a general explosion of R&D in U.S. business and wide business interest in war-generated technologies. See Kim McQuaid, *Uneasy Partners: Big Business in American Politics, 1945–1990* (Baltimore: Johns Hopkins University Press, 1994), chap. 2.

11. Jacob Neufeld, *The Development of Ballistic Missiles in the United States Air Force, 1945–1960* (Washington, D.C.: Office of Air Force History of the U.S. Air Force, 1990), 28–29.

12. Constance McLaughlin Green and Milton Lomask, *Vanguard, a History* (Washington, D.C.: Smithsonian Institution Press, 1971).

13. Edward T. Thompson, "The Rocketing Fortunes of Thiokol," *Fortune* 57, no. 6(June 1954):106–9, 190ff.

14. Virginia P. Dawson, "Building Space Hardware: Industry, National Security, and NASA," in *Space: Discovery and Exploration*, Martin J. Collins and Sylvia K. Kraemer, eds. (Washington, D.C.: Hugh Lauter Levin for the Smithsonian Institution, 1993), 169–219.

15. Michael H. Armacost, *The Politics of Weapons Innovation: The Thor-Jupiter Controversy* (New York: Columbia University Press, 1969).

16. Roger E. Bilstein, *Stages to Saturn: A Technological History of the Apollo/Saturn Launch Vehicle* (Washington D.C.: NASA Scientific and Technical Information Branch, 1980), 14–15. Stuhlinger and Ordway, *Wernher von Braun*, chap. 5.

17. See Neufeld, *Development of Ballistic Missiles*, and Edmund Beard, *Developing the ICBM: A Study in Bureaucratic Politics* (New York: Columbia University Press, 1976).

18. Paul B. Stares, *The Militarization of Space: U.S. Policy, 1945–1984* (Ithaca, N.Y.: Cornell University Press, 1985), 30–33; Jeffrey T. Richelson, *Amer-*

ica's Secret Eyes in Space: The U.S. Keyhole Spy Satellite Program (New York: Harper & Row, 1990)(I am indebted to Dwayne A. Day for this reference); Armacost, *Politics of Weapons Innovation*, chap. 2.

19. Green and Lomask, *Vanguard*. Dwayne A. Day, "A Strategy for Space," *Spaceflight* 38(September 1996):308–12. R. Cargill-Hall, "The Eisenhower Administration and the Cold War: Framing American Astronautics to Serve National Security," in *Organizing for the Use of Space: Historical Perspectives on a Persistent Issue*, Roger D. Launius, ed., AAS History Series, Vol. 18 (San Diego: Univelt, 1995), 49–61.

20. "Giant of Space Industry," *Business Week* (December 23, 1961):100–102ff. Interview with Atwood by Collins. The Navaho also had a ramjet engine, built by Curtiss-Wright.

21. Green and Lomask, *Vanguard*.

22. William B. Harwood, *Raise Heaven and Earth: The Story of Martin Marietta People and Their Pioneering Achievements* (New York: Simon & Schuster, 1993), 231–326, quotation on 288.

23. Rae, *Climb to Greatness*.

24. "A Chronology of LMSC History," courtesy of Beverly Handleman, Lockheed Missile and Space Company, Sunnyvale, Calif., 1–7. Stares, *Militarization of Space*, 31. Robert Kargon, Stuart W. Leslie, and Erica Schoenberger, "Far Beyond Big Science: Science Regions and the Organization of Research and Development," in *Big Science: The Growth of Large-Scale Research*, Peter Galison and Bruce Hevly, eds. (Stanford: Stanford University Press, 1992), 346–51. Rae, *Climb to Greatness*, 199.

25. "Integration: New Force in Rocket Research," *Chemical Week* 77(August 13, 1955):46.

26. "Grand Central Rocket," *Fortune* 54(August 1956):180.

27. David A. Anderson, "Bell Builds Rocket Knowhow," *Aviation Week* 62(January 10, 1955):28.

28. "GE Finds a Profitable Place in Space," *Business Week* (February 18, 1961):126–32.

29. Conversation with Ted D. Smith, September 6, 1994.

30. "Lockheed Missile Scientists Quit," *Aviation Week* 63(December 19, 1955):16.

31. J. Sterling Livingston, "Weapon System Contracting," *Harvard Business Review* 37(July-August 1959):83–92; Robert F. S. Homann, "Weapons System Concepts and Their Pattern in Procurement," *Federal Bar Journal* 17(1957):402–19.

32. Livingston, "Weapon System Contracting."

33. Claude Witze, "Guided Missile Program Reaches 'Pay-Off,'" *Aviation Week* 62(January 31, 1955):13; Draft of a memorandum, General Bernard A. Schriever to Lt. General Thomas S. Power, "Airframe Industries vs. Air Force ICBM Management," December 1954, courtesy of Jacob Neufeld, Department of the Air Force, History Office, Bolling Air Force Base.

34. J. Leland Atwood, conversation with the author, January 10, 1994; Interview with Atwood by Collins.

35. Witze, "Guided Missile Program."

36. Neufeld, *Development of Ballistic Missiles*, chap. 5; Dawson, "Building Space Hardware," 186–88. Convair was the product of a 1943 merger of two pre-war firms, Consolidated Aircraft Corp. and Vultee Aircraft Corp. It was purchased by General Dynamics in 1953.

37. Neufeld, *Development of Ballistic Missiles,* chap. 5 and p. 132.

38. "Survey of Certain Aspects of the Ballistic Missile Program of the Department of the Air Force, as developed by the subcommittee on Manpower Utilization of the Committee on Post Office and Civil Service (House of Representatives) and by the Comptroller General of the United States," (Washington, D.C.: GPO, December 30, 1960).

39. "Survey of Certain Aspects," 25.

40. These paragraphs are based upon chapters 4 and 5 of Green and Lomask, *Vanguard.*

41. Bilstein, *Stages to Saturn*, 15.

42. Koppes, *JPL*, 51–55.

43. Howard E. McCurdy, *Inside NASA: High Technology and Organizational Change in the U.S. Space Program* (Baltimore: Johns Hopkins University Press, 1993), 36–38.

44. Author's conversations with Willis M. Hawkins on April 29 and September 7, 1994, and with Ted D. Smith on September 6, 1994.

45. Armacost, *Politics of Weapons Innovation.*

46. For examples of this thinking, see Claude Witze, "Army Partisans Rap Aircraft Industry," *Aviation Week*, 67(December 9, 1957):26; Rae, *Climb to Greatness*, lays out many of the connections and the antagonisms between the two industries.

47. "Gross Warns Airframe Industry It Is Losing Weapon Business," *Aviation Week* 62(March 28, 1955):14.

48. Van Dyke, *Pride and Power.*

49. Charles D. Bright, *The Jet Makers: The Aerospace Industry from 1945 to 1972* (Lawrence: Regents Press of Kansas, 1978), chap. 1; Roger E. Bilstein, *Flight in America, 1900–1983: From the Wrights to the Astronauts* (Baltimore: Johns Hopkins University Press, 1984).

50. Alex Roland, *Model Research: The National Advisory Committee for Aeronautics, 1915–1958*, NASA History Series, Vol. 1, 205 and 219, and Vol. 2 (Washington, D.C.: NASA, Scientific and Technical Branch, 1985), 684–90.

51. Roland, *Model Research*, Vol. 2, 686–90. Virginia P. Dawson, *Engines and Innovation: Lewis Laboratory and American Propulsion Technology* (Washington, D.C.: NASA, 1991).

52. Roland, *Model Research*, Vol. 1, 186–94 and 199–207.

53. Ibid., Vol. 2, 693–95.

54. Richard Hallion, *Supersonic Flights: The Story of the Bell X-1 and Douglas D-558* (New York: Macmillan, 1972), chap. 2 and 3. Arthur Pearcy, *Flying the Frontiers: NACA and NASA Experimental Aircraft* (Annapolis: Naval Institute Press, 1993).

55. Roland, *Model Research*; John V. Becker, "The X-15 Project," Part 1,

"Origins and Research Background," *Astronautics and Aeronautics* 2(February 1964):52–61. Testimony of Hugh L. Dryden, in "Astronautics and Space Exploration," Hearings before the House Select Committee on Astronautics and Space Exploration, 85/2, April 15 through May 12, 1958.

56. Maxwell W. Hunter, conversation with the author, April 12, 1994, and Charles Feltz, conversation with the author, April 23, 1994.

57. Loyd S. Swenson, Jr., James M. Grimwood, and Charles C. Alexander, *This New Ocean: A History of Project Mercury* (Washington D.C.: NASA, 1966), 73.

58. "'Farside' Aims at 4,000 Mile Altitude," *Aviation Week* 67(July 22, 1957):29; "U.S. Accelerates Moon Plans," *Aviation Week* 67(November 4, 1957):27.

59. Koppes, *JPL*, 84–85.

60. Van Dyke, *Pride and Power*, chap. 3 and 4.

61. They are called geosynchronous because at this altitude, a satellite rotates at the same rate as the earth and so remains fixed above the same spot on the earth.

62. Van Dyke, *Pride and Power*, chap. 3 and 4. House Select Committee on Astronautics and Space Exploration, Hearings, "Astronautics and Space Exploration," 85th Cong., 2d sess., 1958, testimony of Brig. Gen. H. A. Boushey, USAF, 523.

63. Enid Curtis Bok Schoettle, "The Establishment of NASA," in *Knowledge and Power: Essays on Science and Government*, Sanford A. Lakoff, ed. (New York: Free Press, 1966)

64. Swenson, Grimwood, and Alexander, *New Ocean*, 70 and 72–73. House Select Committee on Astronautics and Space Exploration, "Astronautics and Space Exploration," testimony of Arthur Kantrowitz, 509–10.

65. Krafft A. Ehricke, testimony, Senate Committee in Armed Services, "Inquiry into Satellite and Missile Programs," Hearings, 85th Cong., 1st and 2d sess., November 1957–January 1958, 1201–2. Dawson, *Engines and Innovation*, 168–69. The liquid-hydrogen rocket stage was the basis for General Dynamics' Centaur.

66. Senate Committee on Armed Services, Hearings, "Inquiry," 1106. See also the testimonies of Thomas G. Lanphier, Jr., Vice President of Convair-General Dynamics and L. A. Hyland, General Manager of Hughes Aircraft Company.

67. House Select Committee, "Astronautics and Space Exploration," 496.

68. For a comprehensive look at the birth of NASA, see Schoettle, "Establishment of NASA," and Alison Griffith, *The National Aeronautics and Space Act: A Study of the Development of Public Policy* (Washington, D.C.: Public Affairs Press, 1962).

69. The NACA claimed 55 percent of its work was in space in early 1958 (Swenson, Grimwood, and Alexander, *New Ocean*, 84). That was a figure, however, that was easy to massage. See James R. Hansen, *Spaceflight Revolution: NASA Langley Research Center from Sputnik to Apollo* (Washington, D.C.: NASA, 1995), 19–20.

70. Robert A. Divine, *The Sputnik Challenge* (New York: Oxford University Press, 1993), 99–112.

71. House Select Committee, "Astronautics and Space Exploration," 11–15; quotations on 12–13. For the half-million dollar figure see Griffith, *National Aeronautics and Space Act*, 72. The extension of contracting authority to NASA must be seen in the context of a broad trend to more governmental contracting from 1945 to 1960. See Clarence H. Danhof, *Government Contracting and Technological Change* (Washington, D.C.: Brookings Institution, 1968).

72. Swenson, Grimwood, and Alexander, *New Ocean*, 84–85.

73. Quoted from Griffith, *National Aeronautics and Space Act*, 58.

74. "Astronautics and Space Exploration," 1088–90, quotation on p. 1091.

75. Griffith, *National Aeronautics and Space Act*, 71. Senate Committee on Aeronautics & Space Sciences, Hearings, "Investigations into Governmental Organization for Space," 1959, 165–66, and 423.

76. Robert K. Roney, Paraphrase of a conversation, January 11, 1994.

77. Michael Yaffee, "Space Flight May be $4 Billion Market," *Aviation Week* 69(22 September 1958):26.

78. Hallion, *Supersonic Flight*, 34.

79. *HughesNews* (an in-house publication of Hughes Aircraft Company) reported on July 8, 1960, that the ratio went from 1:11 in 1950 to 1:4 in 1960 (p. 5). See also *Aviation Facts and Figures*, 1955, 1957, 1958, and 1959, and Stekler, *Aerospace Industry*, and Baldwin, *Defense Market*.

80. Yaffee, "Space Flight," *Aviation Week* 69(September 22, 1958): 26–27.

81. More generally, the move from engineers or similar corporate leaders to accountants, lawyers, and MBAs was going on throughout U.S. industry. See McQuaid, *Uneasy Partners*, 92.

82. Griffith, *National Aeronautics and Space Act*; Divine, *Sputnik Challenge*, 145–49.

83. Koppes, *JPL*, 102.

84. Roger D. Launius, introduction to *The Birth of NASA: The Diary of T. Keith Glennan*, J. D. Hunley, ed. (Washington, D.C.: NASA, 1993).

85. T. Keith Glennan to Dryden, et al., "Opening Remarks for NASA-Industry Conference," Folder 003443, NASA Historical Reference Collection, NASA Headquarters, Washington, D.C., July 5, 1960.

86. Edward H. Kolcum, "NASA Re-emphasizes Role of Contractor," *Aviation Week* 73(October 3, 1960):32; Ralph J. Cordiner, "Competitive Private Enterprise in Space," Folder 10940, NASA Historical Reference Collection, NASA Headquarters, Washington, D.C., May 4, 1960.

87. John W. Finney, "Space Progress Forcing a Study of Federal Role," *New York Times* (July 25, 1960)1:5. McKinsey and Company, Inc., "An Evaluation of NASA's Contracting Policies, Organization, and Performance," Folder 003419, NASA Historical Reference Collection, NASA Headquarters, Washington, D.C., October 1960.

88. T. Keith Glennan, *The Birth of NASA: The Diary of T. Keith Glennan*, J. D. Hunley, ed. (Washington, D.C.: NASA, 1993), 120. See also Van Dyke, *Pride and Power*, 188–90.

89. Arnold Levine, *Managing NASA in the Apollo Era* (Washington, D.C.: NASA, 1982), 29–33; Robert L. Rosholt, *An Administrative History of NASA, 1958–1963* (Washington, D.C.: NASA, 1966), 154–69. Glennan, *Birth of NASA*.

90. McKinsey and Co., "Evaluation of NASA's Contracting."

91. Ibid., 2–9, 2–11, and 2–12.

92. "Comments to McKinsey Report," submitted by Hans Hueter of Marshall, December 9, 1960, and Memorandum from Harry J. Goett, Goddard Space Flight Center, January 3, 1961, both in Folder 003420, NASA Historical Reference Collection, NASA Headquarters, Washington, D.C.

93. Swenson, Grimwood, and Alexander, *New Ocean*, 91–106. Roland, *Model Research*, chap. 12.

94. Swenson, Grimwood, and Alexander, *New Ocean*, 66–73 and 86–90. Hansen, *Spaceflight Revolution*, 52–57.

95. Swenson, Grimwood, and Alexander, *New Ocean*, 102–5.

96. Hansen, *Spaceflight Revolution*, 37–63.

97. Swenson, Grimwood, and Alexander, *New Ocean*, 138.

98. "NASA Asks Specific Capsule," *Aviation Week* 69(November 24, 1958): 28.

99. Swenson, Grimwood, and Alexander, *New Ocean*, 173; Conversation with Max Faget, January 30, 1995.

Chapter 3 A Tale of Two Companies

1. Appendix E-1, "Space Activities of the U.S. Government," in NASA, *Aeronautics and Space Report of the President, 1989–1990 Activities*, 161. These figures leave out the large secret Defense and intelligence projects.

2. "Rocket Successes Bring Closer Day of Industry in Space," *Business Week* (October 16, 1960):120–22.

3. Donald C. Elder, *Out from Behind the Eight Ball: A History of Project Echo*, American Astronautical Society History Series, Vol. 16 (San Diego: Univelt, 1995), 52–56; Edgar W. Morse, "Preliminary History of the Origins of Project Syncom," *Note* no. 44, NASA Historical Reference Collection, NASA Headquarters History Office, Washington, D.C., September 1, 1964.

4. *Moody's Industrial Manual—1960*. Edward T. Thompson, "The Upheaval at Philco," *Fortune* 59(February 1959):113–15. For the founding of RCA, see Hugh G. J. Aitken, *The Continuous Wave: Technology and American Radio, 1900–1932* (Princeton, N.J.: Princeton University Press, 1985), 355–431.

5. *Moody's Industrial Manual—1960*. Walter Guzzardi, Jr., "GE: The Company Astride Two Worlds," in *The Space Industry: America's Newest Giant*, by the editors of *Fortune* (Englewood Cliffs, N.J.: Prentice-Hall, 1962).

6. Gilbert Burck, "Is AT&T Playing It Too Safe?," *Fortune* 62(September 1960):132ff.

7. Elder, *Project Echo*.

8. J. R. Pierce, *The Beginnings of Satellite Communications* (San Francisco: San Francisco Press, 1968).

9. Morse, "Preliminary History." David J. Whalen, "Billion Dollar Technology: A Short Historical Overview of the Origins of Communications Satellite

Technology, 1945–1965," in *Beyond the Ionosphere: Fifty Years of Satellite Communications*, Andrew J. Butrica, ed. (Washington, D.C.: NASA History Office, 1997), 95–127.

10. Whalen, "Billion Dollar Technology," 100–101; Philip J. Klass, "Civil Communications Satellites Studied," *Aviation and Space Technology*, 70(June 22, 1959):189–91; Willis M. Hawkins, conversation with the author, April 29, 1994; "GE Moves Fast for Place in Space," *Business Week* (May 6, 1961):29.

11. Glennan, *Birth of NASA*; *Business Week* for 1961 provides good coverage of the intrafirm rivalry.

12. Peter Cunniffe, "Misreading History: Government Intervention in the Development of Commercial Communications Satellites," Report no. 24, Program in Science and Technology for International Security, MIT (May 1991): 14–20.

13. Morse, "Preliminary History," 19.

14. I owe this point to Jonathan Coopersmith.

15. For the first of these decisions see Roger D. Launius and J. D. Hunley, *An Annotated Bibliography of the Apollo Program*, NASA History Office Monographs in Aerospace History, no. 2, July 1994, chap. 3.

16. Morse, "A Preliminary History."

17. *Business Week* (March 11, 1961):117.

18. Morse, "Preliminary History," 69. He refers, however, only to NASA's knowledge vis-à-vis Hughes', while I generalize his point to private companies in general.

19. Elder, *Project Echo*, 42.

20. Morse, "Preliminary History," 24–25. Cunniffe, *Misreading History*, 31. James R. Hansen, *Spaceflight Revolution*, 178–79.

21. AT&T Vice President Henry T. Killingsworth, quoted in Philip J. Klass, "AT&T Plans Satellite Launch in One Year," *Aviation Week* 73(October 31, 1960):32.

22. These arguments, from Pierce, *Beginnings of Satellite Communications*, 31–33, echo AT&T testimony at the 1961–2 congressional hearings.

23. Roger A. Kvam, "Comsat: The Inevitable Anomaly," in *Knowledge and Power: Essays On Science and Government*, Sanford A. Lakoff, ed. (New York: Free Press, 1966), 271–92; Jonathan F. Galloway, *The Politics and Technology of Satellite Communications* (Lexington, Mass.: Lexington Books, 1972).

24. Cunniffe, *Misreading History*, 25 and 29. For a NASA view of the differences between Telstar and Relay, see Leonard Jaffe, *Communication in Space* (New York: Holt, Rinehart and Winston, 1966). For an AT&T view, see Pierce, *Beginnings of Satellite Communications*.

25. The act is reprinted in Lloyd D. Musolf, ed., *Communications Satellites in Political Orbit* (San Francisco: Chandler, 1968), 116–28.

26. Galloway, *Satellite Communications*, 88–89.

27. Jaffe, *Communication in Space*, 131–32.

28. Whalen, "Billion Dollar Technology," 115–23.

29. Ibid., 126.

30. Ibid., 125.

31. Cunniffe, *Misreading History,* 31.

32. Charles Murray and Catherine Bly Cox, *Apollo: The Race to the Moon* (New York: Simon & Schuster, 1989), chap. 3 and 4; Bilstein, *Stages to Saturn,* 55.

33. Aerospace Industries Association, *Aerospace Facts & Figures 1961.* William L. Baldwin, *The Structure of the Defense Market, 1955–1964* (Durham, N.C.: Duke University Press, 1967), 177.

34. "Giant of Space Industry," *Business Week* (December 23, 1961): 100–102.

35. William S. Reed, "X-15 Objectives Raised to New Limits," *Aviation Week & Space Technology* 75(November 20, 1961):52–54; Reed, "X-15 Contributes to Structures Research," *Aviation Week & Space Technology* 75(November 27, 1961):57ff.

36. Mike Gray, *Angle of Attack: Harrison Storms and the Race to the Moon* (New York: W. W. Norton, 1992), chap. 3.

37. Bilstein, *Stages to Saturn.* I simplify the very complex story he gives there. In May 1961, two different kinds of boosters were under consideration for the moon mission, the Nova, capable of putting a spaceship directly on the moon, and the Saturns, which lacked the thrust to power "direct ascent" and would need to use one of two indirect methods, "earth rendezvous" or "moon rendezvous."

38. Courtney G. Brooks, James M. Grimwood, and Loyd S. Swenson, Jr., *Chariots for Apollo: A History of Manned Lunar Spacecraft* (Washington, D.C.: NASA, 1979).

39. Brooks, Grimwood, and Swenson, *Chariots for Apollo,* 26–29, and 35–38, especially 26. On Faget's determination to choose the in-house design, see Murray and Cox, *Apollo,* 105–7.

40. Brooks, Grimwood, and Swenson, *Chariots for Apollo,* chap. 2.

41. Barton C. Hacker and James M. Grimwood, *On the Shoulders of Titans: A History of Project Gemini* (Washington, D.C.: NASA Scientific and Technical Information Office, 1977), chap. 2.

42. Brooks, Grimwood, and Swenson, *Chariots for Apollo,* 27–29.

43. Interview with Thomas E. Dolan by Ivan Ertel, October 14, 1968, 23, Johnson Space Center, Apollo Interviews Series, Houston. Gray tells a similar story about a North American approach to Hughes Aircraft Company in Gray, *Angle of Attack,* 94.

44. Edward H. Kolcum, "X-15 Personnel Have Big Role in Apollo, "*Aviation Week & Space Technology* 75(December 11, 1961):37.

45. Gray, *Angle of Attack,* 104–5. These are Gray's words, but are based, I assume, on the "numerous interviews" he held with Storms (293).

46. W. Henry Lambright, *Powering Apollo: James E. Webb of NASA* (Baltimore: Johns Hopkins University Press, 1995), 107–8, 180, 182–83; Lambright, "The NASA-Industry-University Nexus: A Critical Alliance in the Development of Space Exploration," in John M. Logsdon, ed., *Exploring the Unknown: Selected Documents in the History of the U.S. Civilian Space Program,* Vol. 2. *External Relationships* (Washington, D.C.: NASA, 1996), 419.

47. Evert Clark, "North American to Build Apollo Spacecraft," *Aviation*

Week & Space Technology 75(December 4, 1961):26. Lambright, "NASA-Industry-University Nexus," 427, gives the total worth of the first two contracts through 1969 as $4.6 billion, far larger than the awards to any other company.

48. The orbit controversy is covered in a number of books. See Bilstein, *Stages to Saturn*, 60–68, and Murray and Cox, *Apollo*, chap. 7–9.

49. Gray, *Angle of Attack*, 143. Edmund H. Kolcum, "Lunar Bug Contract Due by Mid-October," *Aviation Week & Space Technology*, 77, No. 5 (July 30, 1962):25.

50. *Aviation Week & Space Technology* 75(December 4, 1961):26. Different Saturns were constituted of different stages.

51. James E. Webb, testimony, House Committee on Science and Astronautics, Subcommittee on NASA Oversight, "Investigation into Apollo 204 Accident," hearings, 90th Cong., 1st sess., April 10 . . . , 1967, 7.

52. John M. Logsdon, "From Apollo to Shuttle: Policy Making in the Post-Apollo Era," photocopy, NASA Historical Reference Collection, NASA Headquarters, Washington, D.C., Spring 1983, I-1 to I-5.

53. Levine, *Managing NASA*, 95–99; "NASA Strives for Procurement Flexibility," *Aviation Week & Space Technology* 77, no. 1(July 2, 1962):349ff.

54. Lambright, *Powering Apollo*, 98, 121, and 136–37. Walter A. McDougall, . . . *The Heavens and the Earth: A Political History of The Space Age* (New York: Basic Books, 1985), 380–83. Lambright, "NASA-Industry-University Nexus," 417–18.

55. Webb interview, 150, 168. "Procurement Flexibility," *Aviation Week & Space Technology* 77, No. 1(July 2, 1962):349–51. Glenn E. Bugos, "Program Management in the Department of Defense and NASA: . . . ", March 15, 1987, HHN 157, NASA Historical Reference Collection, NASA Headquarters, Washington, D.C., House Committee, "Apollo 204 Accident," 6.

56. "Report to the President on Government Contracting for Research and Development," Office of the White House Press Secretary, April 30, 1962. In attributing its views to Webb, one of the coauthors, I am making use of Webb's assertion, quoted in Levine, *Managing NASA*, 77, that he played a decisive role in the "important things." On the "Report," see Clarence H. Danhof, *Government Contracting and Technological Change* (Washington, D.C.: Brookings, 1968), chap. 3.

57. Webb interview, 166. Webb specifically attacks the Huntsville practice of designing and manufacturing the first exemplars in-house.

58. Ibid., 187.

59. These numbers are calculated from table 5–1 in Levine, *Managing NASA*, 116.

60. Conversation with J. Leland Atwood, January 10, 1994. Letter from Atwood to the author, January 30, 1995. In Atwood's view, "the things most poorly understood [by NASA monitors] were the techniques of heavy accurate tooling, welding and expansion-contraction problems, structural design and weight reduction activities, and related work."

61. John W. Paup, interviewed by Ivan D. Ertel, June 7, 1966, Apollo Interview Files, Johnson Space Center History Office, Houston, 11.

62. Lambright, *Powering Apollo*, deals in large part with Webb's efforts to keep Apollo going. See also Lambright, "NASA-Industry-University Nexus," and McDougall, *Heavens and the Earth*, chap. 18.

63. One example is given by Gray, *Angle of Attack*, 139–41.

64. See "Grumman May Produce Two Dozen Lunar Landing Modules for NASA," *Aviation Week & Space Technology* 78(January 14, 1963):39; "Apollo Boilerplate Test Program Cut," *Aviation Week & Space Technology* 78(January 28, 1963):58, "LEM Subcontracting Advanced," *Aviation Week & Space Technology* 78(February 4, 1963): 37, and "NASA Asks $800 million for Construction," *Aviation Week & Space Technology* 78(February 25, 1963):37.

65. Bilstein, *Stages to Saturn*, chap. 7.

66. Gray, *Angle of Attack*, 89 and 100–101.

67. Herbert Solow, "North American: A Corporation Deeply Committed," *Fortune* 65, no. 6(June 1962):144–46. "Pure Gold in Space Work," *Business Week* (September 14, 1964):129.

68. Gray, *Angle of Attack*, 78. Solow, "North American."

69. Lambright, *Powering Apollo*, 114–16.

70. Henry C. Dethloff, *Suddenly Tomorrow Came . . . A History of the Johnson Space Center* (Washington, D.C.: NASA Lyndon B. Johnson Space Center, 1993), 70. Gray, *Angle of Attack*, 136.

71. Brooks, Grimwood, and Swenson, *Chariots for Apollo,* 168–69.

72. Gray, *Angle of Attack*, 179–82, quotation on 181.

73. Bilstein, *Stages to Saturn*, 222–25.

74. Brooks, Grimwood, and Swenson, *Chariots for Apollo*, 167–72.

75. McDougall, *Heavens and the Earth*, chap. 19.

76. Levine, *Managing NASA*, 122. Phillips' tiger team is treated extensively in most books on Apollo.

77. Webb interview, 69.

78. Memorandum from S. C. Phillips to G. E. Mueller, December 18, 1965, "CSM and S-II Review," November 15, 1990, in J. Leland Atwood, Apollo File, Oral History Working File, Department of Space History, National Air and Space Museum, Washington, D.C., quotations from pages 1, 3, and 5 of memorandum.

79. J. Leland Atwood, Apollo File. Atwood to Robert C. Seamans, December 30, 1994 (this letter was kindly furnished to me by Mr. Atwood).

80. Bilstein, *Stages to Saturn*, 219–22, quotation on 219. For another example of this technique of joint teams, see Murray and Cox, *Apollo*, 148–51.

81. Lambright, "NASA-Industry-University Nexus," 428. Bilstein, *Stages to Saturn*, 228–30. Brooks, Grimwood, and Swenson, *Chariots for Apollo*, chap. 8. Murray and Cox, *Apollo*, 182–84.

82. Solow, "Deeply Committed," "Earthling Wanted," *Forbes* 98, no. 3(September 1966):44–45. John Mecklin, "The Rockwells Take Off for Outer Space," *Fortune* 75, no. 6(June 1, 1967):100ff.

83. Senate Committee on Aeronautical and Space Sciences, "Apollo Accident," Hearings, April 13 and 17, 1967, 90th Cong., 1st sess., passim. Murray and Cox, *Apollo*, 190–91, 212–15. Brooks, Grimwood, and Swenson, *Chariots for Apollo*, 218–27.

84. For a short discussion of arguments for and against the oxygen atmosphere, see Murray and Cox, *Apollo*, 206-7.

85. For Atwood's perspective, see his pamphlet, "The Apollo Fire," October 1988, in the Oral History Working File, Department of Space History, National Air and Space Museum, Washington, D.C.

86. Lambright, *Powering Apollo*, 170-76. See also Robert C. Seamans, Jr., *Aiming at Targets* (Washington, D.C.: NASA, 1996), 144-45.

87. These paragraphs draw on an interview with William B. Bergen by Loyd S. Swenson, Jr., June 21, 1971, and an interview with John P. Healey by Swenson with Ralph Oakley, July 16 and 21, 1970, both in the Johnson Space Center Apollo Series, Houston. They also make use of Brooks, Grimwood, and Swenson, *Chariots for Apollo*, chap. 9, and Bilstein, *Stages to Saturn*, chap. 7.

88. Bilstein, *Stages to Saturn*, 232. Murray and Cox, *Apollo*, 222.

89. Bergen interview.

90. Brooks, Grimwood, and Swenson, *Chariots for Apollo*, 232 and 253.

91. William H. Becker is among those who have called attention to "the important nexus among private, semi-public and public institutions that has produced advanced technologies." See "Presidential Address: Managerial Culture and the American Political Economy," *Business and Economic History* 25, no. 1(Fall 1996):6.

92. See, for example, "A Vital Year in Space," editorial, *Aviation Week & Space Technology* 82(January 18, 1965):17.

93. Gray, *Angle of Attack*, chap. 10.

94. Bilstein, *Stages to Saturn*, 232.

95. J. Leland Atwood, *Apollo Fire*, SP 20-22.

96. John Paup, interview, Johnson Space Center Archives, Houston.

97. Conversations with Victor G. Savikas, April 28, 1994, and Albert D. Wheelon, January 15, 1994. See also *Federal Reporter*, 2d ser., 463 (1972): 1391.

Chapter 4 The Space Shuttle

1. Robert A. Divine, "Lyndon B. Johnson and the Politics of Space," in *The Johnson Years*, Vol. 2, *Vietnam, the Environment, and Science*, Robert A. Divine, ed. (Lawrence: University Press of Kansas, 1987), 217-53. John M. Logsdon, "From Apollo to Shuttle."

2. Roger D. Launius, "A Waning of Technocratic Faith: NASA and the Politics of the Space Shuttle Decision," *Journal of the British Interplanetary Society* 49(1996):49-58.

3. Competition with the Soviets remained important, but the theater for this competition shifted from NASA to the Defense Department, according to Erasmus H. Kloman, *NASA: The Vision and the Reality* (Washington, D.C.: National Academy of Public Administration, 1985), 3.

4. Bibliographies on all aspects of the shuttle are in Roger D. Launius, "Toward an Understanding of the Space Shuttle: A Historiographical Essay," *Air Power History* 39(Winter 1992):3-18, and Roger D. Launius and Aaron K. Gillette, *Toward A History of the Space Shuttle: An Annotated Bibliography*,

Studies in Aerospace History no. 1 (Washington D.C.: NASA History Office, December 1992).

5. Logsdon, "Apollo to Shuttle," chap. 4. John Logsdon, "The Space Shuttle Decision: Technology and Political Choice," *Journal of Contemporary Business* 7(1978):13. John F. Guilmartin, Jr., and John Walker Mauer, *A Shuttle Chronology, 1964-1973*, JSC 23309 (Houston: NASA, Johnson Space Center, 1988), chap. 3. Douglas R. Lord, "Space Station Program Plans," Briefing to ESRO, June 3, 1970, File 009179, NASA Historical Reference Collection, NASA Headquarters, Washington, D.C.

6. William J. Normyle, "Manned Mission to Mars Opposed," *Aviation Week & Space Technology* 91(August 18, 1969):16. Ken Hechler, *The Endless Space Frontier: A History of the House Committee on Science and Astronautics, 1959-1978*, abridged by Albert E. Eastman, American Astronautical Society History Series (San Diego: Univelt, 1982), chap. 8.

7. Roger D. Launius, *NASA: A History of the U.S. Civil Space Program* (Malabar, Fla.: Krieger, 1994), 216-21, quotation on 218, and 93-98 and 107-10. William J. Normyle, "House Leaders Concur on NASA Budget," *Aviation Week & Space Technology* 92(March 30, 1970):22.

8. Guilmartin and Mauer, *Shuttle Chronology*, chap. 4.

9. The same is true of studies on configurations for space stations.

10. "X-20 Will Probe Piloted Lifting Re-Entry," *Aviation Week & Space Technology* 79(July 22, 1963):230ff; William E. Burrows, "The Military in Space: Securing the High Ground," in *Space: Discovery and Exploration*, Martin J. Collins and Sylvia K. Kraemer, eds. (Hong Kong: Hugh Lauter Levin, 1993), 145.

11. Guilmartin and Mauer, *Shuttle Chronology*, chap. 1; A. Dan Schnyer, interview by John W. Mauer, October 20, 1984, NASA, Johnson Space Center History Office, Space Transportation System Interview Series, Houston.

12. This account is based upon conversations with Maxwell W. Hunter, April 12, 1994, John F. Milton, February 13, 1995, John T. Lloyd, February 24 and March 15, 1994, and Grover L. Alexander, May 20, 1995.

13. M. W. Hunter, W. E. Matheson, and R. F. Trapp. "The Potential of Nuclear Space Transport Systems," Douglas Report No. SM-37427, March 1960, excerpted in "The Origins of the Shuttle (According to Hunter)," Lockheed Missiles & Space Co., September 1972, private files of M. W. Hunter and NASA Historical Reference Collection, NASA Headquarters, Washington, D.C.

14. Memorandum, M. W. Hunter to E. P. Wheaton, "Orbital Transportation," October 28, 1965, private files of M. W. Hunter and NASA Historical Reference Collection, NASA Headquarters, Washington, D.C.

15. Hunter to Wheaton, cited previously. E. P. Wheaton with M. W. Hunter, "Space Commerce," paper presented at the Fourth Goddard Memorial Symposium, American Astronautical Society, Washington, D.C., March 15-16, 1966, private files of M. W. Hunter and NASA Historical Reference Collection, NASA Headquarters, Washington, D.C.

16. Guilmartin and Mauer, *Shuttle Chronology*, I-22-25, I-68-70, I-72-78, II-17-25.

17. Conversation with M. W. Hunter, II, April 12, 1994.

18. Director of Engineering and Development to Director, April 16, 1968, Reading Files of M. A. Faget, Johnson Space Center History Archives, Center Series, Houston.

19. Conversation with Alexander. Dornberger's warning is discussed in chapter 2.

20. Guilmartin and Mauer, *Shuttle Chronology*, chap. 1. Logsdon, "Apollo to Shuttle," chap. 2. George E. Mueller, "The New Future for Manned Spacecraft Developments," *Astronautics and Aeronautics*, 7, no. 3(March 1969):24–32.

21. Alex Roland, "The Shuttle: Triumph or Turkey?" *Discover* 6(November 1985):29–49. Guilmartin and Mauer, *Shuttle Chronology*, chap. 3, pt. 1. Henry C. Dethloff, *Suddenly Tomorrow Came*, 224–25.

22. Max Faget, "Space Shuttle: A New Configuration," *Astronautics & Aeronautics* 8(January 1970):52–61.

23. "An Assessment of The MSC Shuttle Vehicles," August 28, 1969, Box 005–51, Johnson Space Center History Office, Shuttle Program Series, Houston. See also Guilmartin and Mauer, *Shuttle Chronology*, II-150.

24. Guilmartin and Mauer (*Shuttle Chronology*, III-4) write, "Faget . . . borrow[ed] the name from the epoch-making Douglas DC-3 commercial aircraft of the early 1930s, the first land transport to achieve real commercial success."

25. Ibid., III-110–11 and III-116–17. The North American Rockwell contract was part of a group known as Integrated Launch and Reentry Vehicle contracts.

26. Ibid., III-177 to III-182.

27. Ibid., III-198 to III-204, quotation on 199.

28. Ibid., II 17–25 and II 81–95. A Gemini-based shuttle was also of interest to the military for a while.

29. This paragraph and the two that follow are based on ibid., passim.

30. Ibid., IV-1 to IV-10. Conversations with John F. Milton, February 13, 1995, and Grover Alexander, May 20, 1995.

31. Guilmartin and Mauer, *Shuttle Chronology*, IV-19, 59, 153ff.

32. Leroy E. Day Interview by John Mauer, October 17, 1983, Shuttle Oral History Interviews, Johnson Space Center History Office, Houston; and Dale D. Myers, conversations with the author, January 17, and December 12, 1994.

33. Dale D. Myers to Eberhard Rees, Director, Marshall Space Flight Center, Huntsville, June 17, 1970, Drawer 24, "Space Shuttle," folder "Space Shuttle 1970," History Archives, Marshall Space Flight Center, Huntsville. An identical letter was sent to MSC Director Robert Gilruth.

34. E. Rees, 6/14/1970, to Bill Lucas, Drawer 3, File "Space Station #2, 1969–1970," History Archives, Marshall Space Flight Center, Huntsville.

35. Dale D. Myers, Conversations with the author, January 17 and December 12, 1994.

36. Myers, Conversations. Guilmartin and Mauer, *Shuttle Chronology*, V-20.

37. Day, Interview, 40.

38. James B. Jackson, Interview by John F. Guilmartin, Jr., and John W. Mauer, August 5, 1985. Shuttle Oral History Series, Johnson Space Center History Office, Houston.

39. Michael A. Genovese, *The Nixon Presidency: Power and Politics in Turbulent Times* (New York: Greenwood Press, 1990), 61–71.

40. John M. Logsdon, "The Space Shuttle Program: A Policy Failure?" *Science* 232(May 30, 1986):1099–105.

41. Logsdon, "Policy Failure?", quotation on 1100.

42. Willis Shapley, Interview by John W. Mauer, October 26, 1984, Shuttle Oral History Interviews, Johnson Space Center History Office, Houston.

43. In fact, many at NASA had felt that a fully reusable shuttle was unwise for engineering reasons. For example, it would have been difficult to optimize a single engine for both a reusable booster and a reusable orbiter. Guilmartin and Mauer, *Shuttle Chronology*, chap. 4, pt. 1, and Charles J. Donlan, interview by John W. Mauer, October 19, 1983, Shuttle Oral History Interviews, Johnson Space Center History Office, Houston, 19, 20.

44. Guilmartin and Mauer, *Shuttle Chronology*, chap. 5, pt. 1, quotation on V-3.

45. William S. Hieronymous, "Key Shuttle Decisions Near," *Aviation Week & Space Technology* 94(March 29, 1971):45.

46. Michael L. Yaffee, "Program Changes Boost Grumman Shuttle," *Aviation Week & Space Technology* 95(July 12, 1971):37.

47. Zack Strickland, "Titan 3L Studied as Expendable Booster," *Aviation Week & Space Technology*, 95(August 2, 1971):40–41; Zack Strickland, "Winged Saturn Studied for Shuttle," *Aviation Week & Space Technology* 95(September 20, 1971):16; NASA Historical Reference Collection, NASA Headquarters, Washington, D.C., file "Fletcher Correspondence 1971," Memorandum, Administrator to Associate Administrator for Manned Space Flight, July 28, 1971.

48. Guilmartin and Mauer, *Shuttle Chronology*, V-245 and V-281.

49. Ibid., chap. 5, pt. 1, quotation on V-16.

50. Joseph G. Thibodaux, Interview by John F. Guilmartin, Jr. and John W. Mauer, December 12, 1983, Shuttle Oral History Interviews, Johnson Space Center History Office, Houston.

51. Thibodaux interview.

52. McCurdy (*Inside NASA*, 133–41), discusses one of these transformations, industry's growing expertise. Although the DC-3 was set aside, Faget himself was not marginalized. He and his engineers remained central actors in the fashioning of the final form of the shuttle.

53. House Committee on Science and Technology, "Space Commercialization," Hearings before the Subcommittee on Space Science and Applications, May 3 . . . 1983, 98th Cong., 1st sess., 187.

54. Thomas P. Murphy, "The Manned Orbiting Laboratory Controversy," in *Science, Geopolitics, and Federal Spending*, Thomas P. Murphy (Lexington, Mass.: Heath Lexington Books, 1971), 463–87; Dwayne A. Day, "Invitation to Struggle: The History of Civilian-Military Relations in Space," in *Exploring the Unknown*, Vol. 2, John M. Logsdon, ed. (Washington, D.C.: NASA, 1996), 262–63.

55. Gansler, *The Defense Industry*, 12, 21, 26, 171.

56. Ibid., 32–34; Gilbert Burck, "Famine Years for the Arms Makers," *Fortune* 83, no. 5(May 1971):162ff.

57. Burck, "Famine Years"; V. F. Knutzen, "The Aerospace Industry Today and Tomorrow," *Journal of Commercial Bank Lending*, 54(October 1971):11–19; Harold B. Meyers, "For Lockheed Everything's Coming Up Unk-Unks," *Fortune* 80, no. 2(August 1969):76ff.

58. "Growing Competition Faces U.S.," *Aviation Week & Space Technology* 97(July 10, 1972):12–14; "New Strategies Needed as Competition Costs Grow in Burgeoning Aerospace Market," *Aviation Week & Space Technology* 90(June 2, 1969):84ff; Gansler, *Defense Industry*, 171.

59. Vietor, *Contrived Competition*, 41–47; Theodore H. Moran and David C. Mowery, "Aerospace," reprinted from *Daedalus* 120, no. 4(Fall 1991), in U.S. Congress, Joint Economic Committee, Subcommittee on Technology and National Security, "The Aerospace Industry," Hearings, 102d Cong., 1st and 2d sess., 1991 and 1992 (Washington, D.C.: GPO, 1993), 93–117; "Boeing's Future Changes to Cloudy," *Business Week* (March 28, 1970):124–26; "Even the Douglas Trijet is in Trouble," *Business Week* (July 31, 1971):62–63; "Lockheed Hits Heavy Head Winds," *Business Week* (February 14, 1970):46–48.

60. Stares, *Militarization of Space*, chap. 8.

61. "Rockwell Trims the Giant," *Business Week* (January 10, 1970):33; and "Rockwell Trims North American," *Business Week* (January 31, 1970):112. Eleanor Johnson Tracy, "The Outsider Who Boosted Rockwell Into Orbit," *Fortune* 88, no. 3(September 1973):224ff.

62. "The 'One More Chance' Bomber," *Fortune* 82, No. 1(July 1970):27; "Another Rockwell for North American," *Business Week* (November 4, 1972):25.

63. "General Dynamics: in Trouble Again," *Business Week* (October 4, 1969):48–50; "The Real Life Drama at General Dynamics," *Business Week* (July 11, 1970):17; Carol J. Loomis, "Who Wants General Dynamics? Henry Crown, That's Who" *Fortune* 81, no. 6(June 1970):76ff.

64. *Business Week* (1969, 1970, 1971, 1972):passim; Meyers, "For Lockheed, Everything's Coming Up Unk-Unks," *Fortune* 80, no. 2 (August 1969):76ff.

65. "The Ups and Downs of the Hughes Empire," *Business Week* (February 12, 1972):60; "Martin Turns to an Old Hand," *Business Week* (October 28, 1972); "How TRW is Bucking the Recession," *Business Week* (December 5, 1970):42ff.

66. William S. Hieronymous, "$1–Billion Shuttle Engine Program Seen," *Aviation Week & Space Technology* 94(June 21, 1971):60ff.

67. Katherine Johnsen, "Shuttle Engine Speeded by GAO Decision," *Aviation Week & Space Technology* 96(April 10, 1972):14–15.

68. Guilmartin and Mauer, *Shuttle Chronology*, VI-20.

69. Ibid., chap. 6, pt. 1.

70. Claude E. Barfield, "Technology Report/NASA Broadens Defense of Space Shuttle to Counter Critics' Attacks," *National Journal* 4(19 August 1972): 1326; John M. Logsdon, "Choosing Big Technologies: Examples from the U.S. Space Program," *History and Technology* 9(1992):139–50. On the factors that induce politicians to support R&D projects, see Linda R. Cohen and Roger G.

Noll, eds., *The Technology Pork Barrel* (Washington, D.C.: Brookings Institution, 1991).

71. George M. Low, Memorandum for Honorable Jonathan C. Rose, August 14, 1972, enclosure with August 15, 1972, "note for Dr. Fletcher," "Fletcher Correspondence 1972," NASA Historical Reference Collection, NASA Headquarters, Washington, D.C.

72. "Splitting up the Space Shuttle Money," *Business Week* (August 5, 1972): 64.

73. "Space Shuttle Orbiter Subcontractors Chosen," *Aviation Week & Space Technology* 98(April 2, 1973):14. "The Shuttle: Keeping the U.S. Team in Space," *Business Week* (July 27, 1974):52I.

74. Harwood, *Raise Heaven and Earth*, 420–21.

75. "Space Shuttle Solid Rocket Motor Development Awarded to Thiokol," *Aviation Week & Space Technology* 99(November 26, 1973):27; "NASA Restudying Solid Motor Award," *Aviation Week & Space Technology* 101(July 1, 1974):18–19; James C. Fletcher to Frank E. Moss, January 12, 1973, and February 23, 1973, Frank Moss papers, University of Utah libraries (copies courtesy of Lee Saegesser, NASA History Office); Joseph J. Trento with Susan B. Trento, *Prescription for Disaster: From the Glory of Apollo to the Betrayal of the Shuttle* (New York: Crown, 1987), 115–16.

76. Conversations with Grover L. Alexander, May 20, 1995, and John T. Lloyd, February 24, 1994; William B. Schramm, Ronald P. Banas, and Y. Douglas Izu, "Space Shuttle Tile—The Early Lockheed Years," *Lockheed Horizons* 13(1983):12–13.

77. Jeffrey S. Banks, "The Space Shuttle," in Cohen and Noll, eds., *Technology Pork Barrel*, 212.

78. See, for example, W. L. Baldwin, *Structure of the Defense Market*, 116–17.

79. For Nixon's reasons for approving the shuttle, NASA's rationales, and Congress' interests, see Logsdon, "Policy Failure?" and Claude E. Barfield, "Technology Report/NASA Broadens Defense of Space Shuttle to Counter Critics' Attacks," *National Journal* 4(19 August 1972):1323–32.

80. "Intelsat Launches from Shuttle Planned" (*Aviation Week & Space Technology* 105[September 27, 1976]:45) gives fifteen million dollars for a Shuttle launch of 2,000 pounds into geosynchronous orbit, versus twenty-five million dollars for a launch on an Atlas-Centaur rocket, but estimates varied wildly. See House Committee on Science and Technology, "Space Industrialization": Hearing, September 29, 1977, 95th Cong., 1st sess., testimony of George Jeffs; and Barry Miller, "Satcom Orbiting from Shuttle Studied," *Aviation Week & Space Technology* 101(October 21, 1974):55.

81. NASA and the Air Force also bought parts of rockets from each other. See William R. Corliss, *History of the Delta Launch Vehicle*, unpublished manuscript, NASA Historical Reference Collection, NASA Headquarters, Washington, D.C., NASA History Office file 010246, "Delta Documentation (1959–1972)."

82. Conversation with Louis Raburn, February 5, 1995. Corliss, *History of*

the Delta, comments on NASA's zeal in marketing the Delta, but does not try to apportion that zeal between NASA and McDonnell Douglas.

83. Corliss, *History of the Delta,* 4–13 to 4–16 and 5–2. This kind of situation, in which companies compete with each other through competition between the government agencies with which they are affiliated, is endemic.

84. "Project Approval Document: Research and Development: Hardware Development/Flight," September 26, 1978, folder "Launch Vehicles, General (1968–1978)," NASA Historical Reference Collection, NASA Headquarters, Washington, D.C.

85. Logsdon, "Policy Failure?"; "Pro and Con Discussion: Should the Proposed Future Manned Space Flight Programs Be Curtailed?" *Congressional Digest* 51(June–July 1972):161–92; Craig Covault, "GAO to Suggest Further Shuttle Delay," *Aviation Week & Space Technology* 104(March 1, 1976):22–23.

86. "NASA Planetary Offices Plans Appeal on Interim Upper Stage," *Aviation Week & Space Technology* 102(May 26, 1975):19; Donald E. Fink, "Upper Stage Plans for Shuttle Detailed," *Aviation Week & Space Technology* 103(October 27, 1975):14; Katherine Johnsen, "Spin-Stabilized Stage for Shuttle Studied," *Aviation Week & Space Technology* 104, no. 22(May 31, 1976):61. Other companies also bid.

87. Barry Miller, "NASA Sets Briefing on Upper Stages," *Aviation Week & Space Technology* 105(September 6, 1976):46. See also Trento, *Prescription for Disaster,* 170–72.

88. Charles A. Ordahl, McDonnell Douglas Astronautics, testimony, House Committee on Science and Technology, "Space Commercialization," 83–85.

89. Trento, *Prescription for Disaster,* 170–72.

90. Robert J. Naumann and Harvey W. Herring, *Materials Processing in Space: Early Experiments* (Washington, D.C.: NASA, 1980).

91. Senate Committee on Aeronautical and Space Sciences, "Space Shuttle Payloads," Hearing, Part I, 93d Cong., 1st sess., October 30, 1973.

92. Erwin J. Bulban, "Non-Aerospace Firms Eyed for Shuttle," *Aviation Week & Space Technology* 104(April 5, 1976):52.

93. Robert Hotz, "A New Space Era," *Aviation Week & Space Technology* 105(October 4, 1976):9. The words are Hotz's, paraphrasing Goldwater.

94. "The Shuttle Opens the Space Frontier to U.S. Industry," *Business Week* (August 22, 1977):48.

95. Marshall Space Flight Center History Archives, Huntsville, Drawer #46, folder "Payload Planning Activity," PD-SA-L, June 1971.

96. "Charter for a Discipline Center for Space Processing Applications," enclosure in W. R. Lucas to NASA Headquarters/Bradford Johnston, July 23, 1976, Drawer #32, folder "Materials Processing in Space, 1976–1977," Marshall Space Flight Center History Archive, Huntsville.

97. "Offer for a Joint Endeavor Between NASA and Microgravity Research Associates, Inc.," received April 9, 1979, NASA Historical Research Collection, file 010937; Edward H. Kolcum, "Company Plans to Manufacture Crystals in Space," *Aviation Week & Space Technology* 120(June 25, 1984):100–101.

98. Peter E. Glaser, "Power from the Sun: Its Future," *Science* 162(Novem-

ber 22, 1968):857–61; Glaser, "History and Outlook," in Peter E. Glaser, Frank P. Davidson, and Katinka I. Csigi, *Solar Power Satellites: The Emerging Energy Option* (New York: Ellis Horwood, 1993).

99. Conversation with Gordon R. Woodcock, May 25, 1995. See Gordon R. Woodcock, "Solar Power from Space: Potential for Large-Scale Enterprise," *Journal of Contemporary Business* 7, no. 3(1978):127–42.

100. Benjamin M. Elson, "Space-Based Solar Power Study Nears Completion," *Aviation Week & Space Technology* 107(September 19, 1977): 58; Craig Covault, "Views Change on Power Satellite Work," *Aviation Week & Space Technology* 109(July 17, 1978):42. Solar satellites were popular with Congress, and there were a series of hearings on them in the late 1970s.

101. House Committee on Science and Technology, "Space Industrialization," 95th Cong., 1st sess., September 29, 1977, testimony of George W. Jeffs, 74ff.

102. Conversation with Charles D. Walker, March 10, 1995.

103. Adam L. Gruen, *The Port Unknown: A History of the Space Station Freedom Program*, manuscript, NASA History Office, April 20, 1993, chap. 2.

104. This is an inference from Walker's remarks.

105. This and the following paragraphs are based on my conversation with Charles D. Walker, March 10, 1995, and on James T. Rose and Terrence D. Fitzpatrick, "The Potential of Materials Processing Using the Space Environment," Ivan Bekey and Daniel Herman, eds., *Space Stations and Space Platforms—Concepts, Design, Infrastructure, and Uses*, Vol. 99 of *Progress in Astronautics and Aeronautics* (New York: American Institute of Aeronautics and Astronautics, 1985), 167–200.

106. Stephen P. Strickland, *Politics, Science and Dread Disease: A Short History of United States Medical Research Policy* (Cambridge, Mass.: Harvard University Press, 1972).

107. Report of the Committee on Scientific and Technological Aspects of Materials Processing in Space of the Space Applications Board, Assembly of Engineering, National Research Council, *Materials Processing in Space* (Washington D.C.: National Academy of Sciences, 1978).

108. John Gilman, "Why Industry Fights Shy of Space," *New Scientist* 132, no. 1790(October 12, 1991):63.

109. House Committee on Science and Technology, "Space Industrialization," Hearing, September 1977, 95th Cong., 1st sess., 91.

110. Barry Miller, "Satcom Orbiting From Shuttle Studied," *Aviation Week & Space Technology* 101(October 21, 1974):55.

111. Maxwell W. Hunter II, Wayne F. Miller, and Robert M. Gray, "The Space Shuttle Will Cut Payload Costs," *Astronautics & Aeronautics* 10(June 1972):50–58. In fact "most satellites launched on the Shuttle require *more* folding, not less. The Shuttle bay is a 15' × 60' cylinder. The typical ELV shroud is a 10' × 10' cylinder with an additional 10' long cone. Use of the Shuttle required that whatever was in the cone now had to be folded." David J. Whalen to the author, letter, December 20, 1996.

112. Bruce A. Smith, "Hughes Seeking Comsat Cost Cuts," *Aviation Week & Space Technology* 112, no. 26(June 30, 1980):54–56.

113. Interview with Joseph McGoldrick by John Mauer, October 24, 1984, Shuttle Interviews, Johnson Space Center History Office, Houston, 34–36.

114. David J. Whelan to the author, December 20, 1996.

115. Albert D. Wheelon, "Trends in Satellite Communications," paper presented to the 4th World Telecommunication Forum, Geneva, Switzerland, October 28, 1983, 22.

116. Wheelon, "The Economics of Telecommunications in the Century of the Satellite," World Telecommunications Forum, Geneva, Switzerland, September 20, 1979, 4.

117. Conversation with Ted D. Smith, September 6, 1994.

118. Wheelon ("Trends in Satellite Communications," 22) gives these figures for launch insurance rates in 1983: 4 to 7 percent for the shuttle versus 5 to 8 percent for Delta and 10 to 14 percent for Ariane.

119. "Intelsat Launches from Shuttle Planned," *Aviation Week & Space Technology* 105, no. 13(September 27, 1976):45.

120. Offsets are demands that, in exchange for a purchase, the selling company or nation will use a certain fraction of the purchase price to buy products from or will build a given portion of a system in the purchasing country.

121. Conversation with George D. Baker, March 13, 1995.

122. NASA History Office, Archives, file 10244, "Delta Documentation (1976–)." See the correspondence between Robert F. Chinnick of Telesat Canada and John F. Yardley, April 13 and May 12, 1976, and "Request for NASA Management Directive Clearance," July 7, 1978.

123. Alex Roland, "The Shuttle: Triumph or Turkey?" *Discover* 6(November 1985):38, 39.

124. In 1975, NASA had projected sixty flights a year by 1984. "Planners Use New Shuttle Traffic Model," *Aviation Week & Space Technology* 102, no. 21(May 26, 1975):49. Craig Covault, "Shuttle Demand Shifts Policies," *Aviation Week & Space Technology* 111, no. 18(October 29, 1979):16.

125. Bruce A. Smith, "Shuttle Launch Delays' Cost Impact Assessed," *Aviation Week & Space Technology* 112, no. 17(April 28, 1980):18–19; Covault, "Demand Shifts Policies."

Chapter 5 Space and the Marketplace

1. For the Ariane launching see Guy Collins, *Europe In Space* (London: Macmillan, 1990), 49–52.

2. Joseph J. Hogan, "Reaganomics and Economic Policy," in *The Reagan Presidency: An Incomplete Revolution?*, Dilys M. Hill, Raymond A. Moore, and Phil Williams, eds. (New York: St. Martin's Press, 1990).

3. U.S. Congress, "Policy and Legal Issues Involved in the Commercialization of Space," Report by the Congressional Research Service for the Senate Committee on Commerce, Science, and Transportation, Sept. 23, 1983 (Washington, D.C.: GPO, 1983).

4. Philip Klutznick, testimony (p. 3), and James Beggs, testimony (p. 26),

House Committee on Science and Technology, "Space Commercialization," Hearings, May 3, 4, 17, 18, 1983, 98th Cong., 1st sess., 3, 26; Adlai E. Stevenson, "The New Era in Space," *Journal of Contemporary Business* 7, no. 3(1978):7–12.

5. Dexter C. Hutchins, "Entrepreneurs Aim for Outer Space," *Venture* 2(September 1980):48ff.

6. Interview with John Klineberg, *Space News* (May 8–14, 1995):30.

7. Daniel Lindley, "No Free Launch: Tracking Space Fledglings," *Barron's* 64(September 17, 1984):47ff. David P. Gump, *Space Enterprise: Beyond NASA*, (New York: Praeger, 1990), 37–39; "Percheron: Private Enterprise in Orbit," *The Miami Herald* (June 8, 1981):11ff. "Private Firms Challenge NASA's Space Monopoly," *The Washington Post* (July 29, 1981):A1ff. "Texas Rocket Built on 'Shoestring' Carries Free Enterprise Into Space," *New York Times*, September 10, 1982, A1. These news articles are in the NASA Historical Reference Collection, NASA Headquarters, Washington, D.C., file 010782, "Private Launchers, SSI." House Committee on Science and Technology, "Future Space Programs: 1981," Hearings, September 21–23, 1981, 97th Cong., 1st sess; Hearings, "Space Commercialization," before the same committee, May 3, 4, 17, 18, 1983, 98th Cong., 1st sess. Space Services Inc. was later acquired by EER Systems.

8. James C. Bennett, "The Second Space Race," *Reason* (November 1981): 21–31.

9. Lindley, "No Free Launch"; Gump, *Beyond NASA*, 28–31; Walter W. Kovalick, Jr., "Space Launch Services: Public or Private Operation?", May 1982, File 010782, NASA Historical Reference Collection, NASA Headquarters, Washington, D.C.

10. Conversation with George Baker, March 13, 1995; "Astrotech Offering Satellite Processing," *Aviation Week & Space Technology* 120(June 25, 1984): 122–23; Letter of Robert J. Goss to the author, February 27, 1996; Robert J. Goss, "Space Commercialization—The Problems and Opportunities," talk at Pittsburgh University Space Symposium, 1985 (private files of R. J. Goss); Don N. Stitt, private communication.

11. Nancy O. Perry, "Shooting for the Stars," *Harvard Business School Bulletin* (June 1989):2–11, quotation on 4 (courtesy of Laura Ayres, Orbital Sciences Corp.).

12. House Committee on Science and Technology, "Space Commercialization," 1983, 93; Perry, "Shooting for the Stars," quotations from Ferguson on 9; "Orbital Sciences Offers Upper Stages," *Aviation Week & Space Technology* 120(June 25, 1984):108ff; Nathan C. Goldman, *Space Commerce: Free Enterprise on the High Frontier* (Cambridge, Mass.: Ballinger, 1985), 49–50.

13. Hutchins, "Entrepreneurs Aim for Outer Space"; "C2 Spacelines Signs for Shuttle Slots, Will Lease Cargo Space," *Aviation Week & Space Technology* 120, no. 26(June 25, 1984):44; James K. La Fleur, "Private Sector Investment in the Space Program: Why, How and When," in "Materials Processing in Space, 1982–1983," Drawer 32, Marshall Space Flight Center History Office Files, Marshall Space Flight Center, Huntsville. (GTI soon got out, however.) House Committee on Science, Space and Technology, "NASA's Commercial Space Program," Hearing, October 20, 1993, 103d Cong., 1st sess., 83–85.

14. Carole A. Shifrin, "Investors Taking Cautious View of Private Programs," *Aviation Week & Space Technology* 120(June 25, 1984):78.

15. See the testimony of David Hannah, in House Committee on Science and Technology, "Future Space Programs: 1981," September 21–23, 1981, 97th Cong., 1st sess., 96, and *New York Times*, September 10, 1982, A1.

16. For a discussion and criticism of this view see Bennett Harrison, *Lean and Mean: The Changing Landscape of Corporate Power in the Age of Flexibility* (New York: Basic Books, 1994).

17. U.S. Senate, Committee on Commerce, Science, and Transportation, "Commercial Space Launch Act," Hearing, September 6, 1984, 98th Cong., 1st sess. (sic), 58.

18. George Gilder, *The Spirit of Enterprise* (New York: Simon and Schuster, 1984).

19. "Investment Firm Unit Considers Private Buy of Space Shuttle," *Aviation Week & Space Technology* 116(January 4, 1982):23.

20. "United States Space Policy: Fact Sheet Outlining the Policy," July 4, 1982, *Weekly Compilation of Presidential Documents* 18, no. 27(July 12, 1982): 872–76.

21. National Academy of Public Administration, "Encouraging Business Ventures in Space Technologies," Report by a Panel of the National Academy of Public Administration, prepared for the National Aeronautics and Space Administration (NASA), May 1983. Copy on file, Library, NASA Headquarters, Washington, D.C.

22. Box "NASA Headquarters Organization," folder, "Office of Commercial Programs"; files of J. M. Beggs, folder "Beggs Correspondence, 1981–1983," folders 010931 and 010932, NASA Historical Reference Collection, NASA Headquarters, Washington, D.C.; Conversation with Isaac T. Gillam, February 20, 1996.

23. Isaac T. Gillam, testimony, House Committee on Science and Technology, "1987 NASA Authorization," Hearings, February to April 1986, 99th Cong., 2d sess., 440–505. Conversation with Gillam. Peter T. Eaton to the author, February 22, 1996, and April 2, 1996. Conversation with James Fountain, November 16, 1994.

24. Memorandum, "Consolidation of NASA Commercial Space Activities," from Administrator [Beggs] to Officials-in-Charge of Headquarters Offices, June 19, 1985 (private files of P. T. Eaton).

25. NASA Advisory Council, "Report of the Task Force for the Commercial Use of Space," August 1, 1985, Files of the Advisory Council, 1984–1986, NASA Historical Reference Collection, NASA Headquarters, Washington, D.C., 10.

26. Isaac T. Gillam, testimony, House Committee on Science and Technology, "1987 NASA Authorizations," quotation on 497. The OCP was merged in 1992 into the Office of Advanced Concepts and Technology.

27. Howard E. McCurdy, *The Space Station Decision: Incremental Politics and Technological Choice* (Baltimore: Johns Hopkins University Press, 1990); Adam L. Gruen, *Port Unknown*.

28. See McCurdy, *Space Station Decision*; Gruen, *Port Unknown*; and House

Committee on Science and Technology, "NASA's Space Station Activities," Hearings, August 2, 1983, 98th Cong., 1st sess., testimony of John Hodge and Peter Wood.

29. Thomas J. Lewin and V. K. Narayanan, *Keeping the Dream Alive: Managing the Space Station Program, 1982–1986* (NASA Contractor Report 4272, July 1990); Sylvia D. Fries, *NASA Engineers and The Age of Apollo* (Washington, D.C.: NASA, 1991), 170–71.

30. See, for example, Robert L. Walquist, TRW, testimony, House Committee on Science and Technology, "1983 NASA Authorizations," Hearings, 974.

31. On lobbying Reagan, see Gruen, *Port Unknown*, 108. On jockeying, see Lewin and Narayanan, *Keeping the Dream Alive*, 45.

32. McCurdy, *Space Station Decision*, pt. 3.

33. Lewin and Narayanan, *Keeping the Dream Alive*, chap. 5 and 6. The exact content of the work packages continued to be negotiated past early 1984.

34. Gerald M. Steinberg, "The Militarization of Space: From Passive Support to Actual Weapons Systems," in *The Exploitation of Space: Policy Trends in the Military and Commercial Uses of Outer Space*, Michiel Schwar and Paul Stares, eds. (London: Butterworth, 1985), 31–49.

35. Stares, *Militarization of Space*, 157–58.

36. Ibid., chap. 8–10.

37. Phil Williams, "The Reagan Administration and Defense Policy," in *The Reagan Presidency*, Hill, Moore, and Williams, eds., 199–230; House Committee on Science and Technology, "National Space Policy," Hearing, August 4, 1982, 97th Cong., 2d sess.

38. Stares, *Militarization of Space*, chap. 11.

39. U.S. Congress, Office of Technology Assessment, *SDI: Technology, Survivability, and Software*, OTA-ISC-353 (Washington, D.C.: GPO, May 1988), 24; Stares, *Militarization of Space*, 227. "Martin Marietta Stresses Technology in Two Areas," *Aviation Week & Space Technology* 116(February 22, 1982):66–67; Conversation with Steven S. Myers, 26 April 1994; Edward Reiss, *The Strategic Defense Initiative* (Cambridge: Cambridge University Press, 1992).

40. Craig Russell Reed, *U.S. Commercial Space Launch Policy Implementation, 1986–1992*, Ph.D. dissertation, George Washington University, Washington, D.C., 1998, 37–42, quotation on 38.

41. Reed, ibid., gives the price in 1983 as thirty-eight million in 1982 dollars and the short-run marginal cost as forty-two million dollars, 40.

42. Jack Scarborough, "The Privatization of Expendable Launch Vehicles: Reconciliation of Conflicting Policy Objectives," *Policy Studies Review* 10(Spring–Summer 1991):12–30, and Senate Committee on Commerce, Science and Transportation, "Commercial Space Launch Act," Hearing, September 6, 1984, 98th Cong., 1st sess. (sic); Reed, *Implementation*, 175–87.

43. See, for example, Beggs' testimony in House Committee on Science and Technology, "Space Commercialization," May 1983. See also Scarborough, "Privatization."

44. Kovalick, "Space Launch Services," 14.

45. Charles A. Ordahl, testimony, House Committee on Science and Technology, "Space Commercialization," 82–84, quotation on 85.

46. Memorandum, "Launch Vehicle Review at Close of GFY '82," T. D. Smith to A. P. O'Neal, October 19, 1982, private files of T. D. Smith. For an analysis of the launch rates and flight costs over the shuttle's first decade, see Roger A. Pielke, Jr., and Radford Byerly, Jr., "The Space Shuttle Program: Performance Versus Promise," in *Space Policy Alternatives,* Radford Byerly, Jr., ed. (Boulder, Colo.: Westview Press, 1992), 223–45.

47. "NASA to Negotiate with Transpace Carriers for Delta," *Defense Daily* (January 10, 1984):35, NASA Historical Reference Collection, NASA Headquarters, Washington, D.C.

48. Edmund L. Andrews, "Lost in Space," *Venture* 8(December 1986):38ff. David W. Grimes, testimony, House Committee on Science and Technology, "1987 NASA Authorization, Vol II," March 20, 1986, 99th Cong., 2d sess., 525–55. James M. Beggs to Don Fuqua, April 10, 1984, in folder 010788, "Transpace Carriers, Inc.," NASA Historical Reference Collection, NASA Headquarters, Washington, D.C.

49. "Space Transportation Firm Agrees to Marketing of Titan," *Aviation Week & Space Technology* 117(December 6, 1982):26–27. Space Transportation Company is discussed in more detail later.

50. Gareth B. Flora and Klaus Heiss, testimony, House Committee on Science and Technology, "Space Commercialization," Hearings, 1983, 160–82, 182–214. "Federal Express Wants to Deliver in Space," *Business Week* (July 4, 1983):42.

51. National Research Council, *Assessment of Candidate Expendable Launch Vehicles for Large Payloads,* 1984, 8, NASA Historical Reference Collection, NASA Headquarters, Washington, D.C. Reed, *Implementation,* 269–70.

52. This and the following paragraphs are based on the testimony of William Rector at the hearings on "Space Commercialization," and at Senate Committee on Science, Technology and Space, "Commercial Space Launch Act," 1984.

53. Testimony of Department of Defense spokesman Herbert A. Reynolds in Hearings, "Space Commercialization," 149ff.

54. House Committee on Appropriations, "Department of Defense Appropriations for 1985, Part 6," Hearings, May 22, 1984, 98th Cong., 2d sess., 645–99; Eugene Kozicharow, "Air Force Developing Plan on Expendable Launching," *Aviation Week & Space Technology* 120(June 18, 1984):70ff; Alton K. Marsh, "Report Warns on Loss of Launcher Production," *Aviation Week & Space Technology* 118(May 23, 1983):23ff; Bruce A. Smith, "Greater Defense Shuttle Role Urged," *Aviation Week & Space Technology* 116(February 1, 1982):22; "Air Force Defends ELV Plan, Cites Shuttle Costs," *Aerospace Daily* 126, no. 17(March 23, 1984):129–30; National Research Council, Committee on NASA Scientific and Technological Program Reviews, "Assessment."

55. House Committee on Science and Technology, "Initiatives to Promote Space Commercialization," Hearings, June 19, 1984, 98th Cong., 2d sess., 17.

56. "NASA Offers Launch Vehicle Using Space Shuttle Elements," *Aviation Week & Space Technology* 120, no. 20(May 14, 1984):18–19.

57. Howard E. McCurdy, *Inside NASA,* 144–46.

58. Donna Blackshear and Herbert Roche, "A Summary of Studies on STS Management Strategy," September 1984, Box 1, File "STS Management Studies," Johnson Space Center History Office, Center Series, Houston, gives the opinions of fifteen studies, conducted from 1977 to 1984.

59. See, for example, Grahame Thompson, *The Political Economy of the New Right* (Boston: Twayne, 1990), and David Steel and David Heald, "The Privatization of Public Enterprises, 1979–1983," in *Implementing Government Policy Initiatives: The Thatcher Administration, 1979–1983,* Peter Jackson, ed. (London: Royal Institute of Public Administration, 1985), 69–91.

60. See, for example, Steve H. Hanke, ed., *Prospects for Privatization* (New York: Academy of Political Science, 1987).

61. Pamela E. Mack, *Viewing the Earth: The Social Construction of the Landsat Satellite System* (Cambridge, Mass.: MIT Press, 1990), chap. 14; Marcia S. Smith, "Civilian Space Applications: The Privatization Battleground," in Radford Byerly, Jr., *Space Policy Reconsidered* (Boulder, Colo.: Westview Press, 1989), 105–16. Congress beat back the attempt to privatize the weather satellites.

62. *Report on Privatization by the President's Private Sector Survey on Cost Control* (Washington D.C.: GPO, 1983), 87.

63. McCurdy, *Inside NASA,* quotation on 144.

64. Pielke and Byerly, "The Space Shuttle Program," 237. In 1985, the shuttle would fly nine missions, the highest annual rate for the decade. Edward H. Kolcum, "NASA Assesses Effects of Failure in Launch of Discovery," *Aviation Week & Space Technology* 121(July 2, 1984):16.

65. "Report of the Shuttle Operations Strategic Planning Group," March 7, 1985, File 007985, NASA Historical Reference Collection, NASA Headquarters, Washington, D.C., quotation on 29.

66. *Insight,* November–December 1981, 8, NASA Historical Reference Collection, NASA Headquarters, Washington, D.C. For the argument that NASA lacked legal grounds for selling commercial launches via the shuttle, see George S. Robinson, "Private Management and Operation of the Space Shuttle: . . . ," *Akron Law Review* 13(Spring 1980):601–11.

67. James Beggs, testimony, House Committee on Science and Technology, "Initiatives to Promote Space Commercialization," June 19, 1984, 98th Cong., 2d sess., 17.

68. Klaus P. Heiss, "A Perspective on Private Initiatives in U.S. Space Transportation," *Journal of Social, Political, and Economic Studies* 11, no. 1(Spring 1986):3–16.

69. House Committee on Science and Technology, "The Need for a Fifth Shuttle Orbiter," Hearings, June 15, 1982, 97th Cong., 2d sess. "NASA Prepared to Consider Private Offer to Fund Fifth Orbiter," *Defense Daily* (January 27, 1982):120.

70. Craig Covault, "NASA Planning for Shift of Shuttle Marketing Operations," *Aviation Week & Space Technology* 117(November 1, 1982):16. "Report of the Shuttle Operations Strategic Planning Group," March 7, 1985, File 00 79

85, NASA Historical Reference Collection, NASA Headquarters, Washington, D.C., 10.

71. Guy Collins, *Europe in Space*, chap. 7, esp. 53–54. Memo, J. F. Yardley to C. J. Dorrenbacher, "Launch Vehicle Sales for NASA," April 26, 1982, private files of Ted D. Smith.

72. Heiss, "Perspective on Private Initiatives," quotation on 7; "Space Transportation Firm Agrees to Marketing of Titan," *Aviation Week & Space Technology* 117, no. 23(December 6, 1982):26–27. "SpaceTran Does Not Need Decision on Fifth Orbiter This Year," *Defense Daily* (March 16, 1983):91. Hearings, "Space Commercialization," 212–13. The details of Heiss's proposal varied with time.

73. Kovalick, "Space Launch Services"; "Managers Find Shuttle Proposal Needs Changes," *Aviation Week & Space Technology* 117, no. 24(December 13, 1982):17.

74. Heiss, "Perspective on Private Initiatives," 9–10; Conversation with Isaac T. Gillam, February 20, 1996.

75. Bill Saporito, "Retirement is a Blast for Al Rockwell," *Fortune* 112 (August 5, 1985):82–84, quotation on 84. See also "Two Firms Ready to Buy, Produce Shuttle Orbiters," *Aviation Week & Space Technology* 120(June 25, 1984): 116ff.

76. Telephone conversation with Don N. Stitt, February 25, 1996; Conversation with Isaac T. Gillam, February 20, 1996. Robert J. Goss, "Space Commercialization—The Problems and Opportunities," presentation at Pittsburgh University, 1985 (private files of R. J. Goss). Matthew Heller, "Al Rockwell's Space Odyssey," *Forbes* 136, no. 16(December 30, 1985):54–55.

77. House Committee on Science and Technology, "The Need for a Fifth Space Shuttle Orbiter," 132.

78. Robert L. Brock of Boeing and William S. Field of Prudential Insurance Company, testimony, U.S. House Committee on Science and Technology, "The Need for a Fifth Space Shuttle Orbiter," 92–112, 120–49.

79. This section is heavily indebted to Peter Cunniffe, *Misreading History*.

80. Conversations with Samuel W. Fordyce, September 27, 1995, and Richard T. Gedney, 18 March 1996. For a history of Lewis' collaborations with the engine industry, see Virginia P. Dawson, *Engines and Innovation*.

81. Cunniffe, *Misreading History,* 73–75.

82. House Committee on Science and Technology, "NASA Space Communications Program," July 8,9, 1981, 97th Cong., 1st sess. "NASA Role in Communications Satellite R&D: White Paper Part of Outlook for Space," Box 222, folder 5.2, "Materials for 30/20 (ACTS) Advocacy," Lewis Research Center Records, Cleveland.

83. Representative Bob Shamansky in "NASA Space Communications Programs," 85.

84. Cunniffe, *Misreading History*, 155–57.

85. See Charles Rosenberg, "Toward an Ecology of Knowledge: On Discipline, Context, and History," in *The Organization of Knowledge in Modern America, 1860–1920,* Aleandra Oleson and John Voss, eds. (Baltimore: Johns Hopkins, 1979); Alex Roland, *Model Research: The National Advisory Commit-*

tee for Aeronautics, 1915–1958 (Washington, D.C.: NASA History Series, 1985). The same negotiation takes place within companies. See, for example, Margaret B. W. Graham, *RCA and the VideoDisc: the Business of Research* (Cambridge: Cambridge University Press, 1986).

86. Cunniffe, *Misreading History,* 153. See also McCurdy, *Inside NASA,* 134–41.

87. Daniel R. Glover, "NASA Experimental Communications Satellites, 1958–1995," in *Beyond the Ionosphere: Fifty Years of Satellite Communication,* Andrew J. Butrica, ed. (Washington, D.C.: NASA History Office, 1997), 51–64.

88. House Committee on Science and Technology, Hearings, "NASA Space Communications Program," 18, 73–76, quotation by RCA executive and Ad Hoc Subcommittee member, John E. Keigler, on 67.

89. Conversation with Samuel W. Fordyce, September 27, 1995.

90. House Committee on Science and Technology, "NASA Space Communications Program"; House Committee on Science and Technology, "Communications Research and Development," Hearings, May 1980, 96th Cong., 2d sess.

91. Arno A. Penzias, testimony, House Committee on Science and Technology, "Communications Research and Development," 205–35.

92. Penzias testimony. Methods were subsequently found to overcome rain fade, but they were expensive.

93. Conversations with Robert K. Roney, September 25, 1995, and Albert D. Wheelon, August 12, 1995. Dr. Roney comments on this sentence, "I believe it is unfair to contrast Hughes' position on ACTS [with] its position in Syncom days. The big unknown . . . in 1959–1963 had to do with spaceflight itself, not communications electronics. That was precisely what NASA was created to explore" (letter to the author, April 7, 1998).

94. See Bruce L. R. Smith, *American Science Policy Since World War II* (Washington, D.C.: Brookings Institution, 1990). Smith claims that Reagan's policy was, however, inconsistent in that it was a jumble of ideological precepts, pragmatic responses to problems, and programs and policies inherited from the Carter administration.

95. Philip J. Klass, "Advanced Satcom Designed for Military," *Aviation Week & Space Technology* 115(September 28, 1981):72; "Keeping Lines Open During a Nuclear War," *Business Week* (February 7, 1983):116–17; Cunniffe, *Misreading History,* 83 and 100–101.

96. Cunniffe, *Misreading History,* 76–87; and Marcia S. Smith, "NASA's Advanced Communications Technology Satellite (ACTS) Program in Light of the Hughes Filing." March 2, 1984, Report of the Congressional Research Service, Library of Congress.

97. "Keeping Lines Open During a Nuclear War," *Business Week* (February 7, 1983):116–17.

98. Chris Bulloch, "Communications Satellite Prospects: Competition Sharpens Between the 'Big Three' U.S. Builders," *Interavia* (October 1983):1111–13, quotation on 1112. *The Wall Street Journal Index* (Princeton, N.J.: Dow Jones Books, 1958–80).

99. Cunniffe, *Misreading History,* 80–85.

100. The quotations are from Cunniffe, *Misreading History,* 82, who took them from Jay C. Lowndes, "RCA Marshalling Support Behind Advanced Satcom," *Aviation Week & Space Technology* 120(January 9, 1984):24–25.

101. Cunniffe, *Misreading History,* 85–90. Conversations with Rod Knight, March 19, 1996, and Fordyce.

102. Ronald J. Schertler, Richard T. Gedney, and Michael J. Zernic, "ACTS' Legacy: The Ka-band Explosion," *Aerospace America* 34(February 1996):33–36.

103. Schertler, Gedney, and Zernic, "ACTS' Legacy"; Patrick Seitz, "Lockheed, Others Join Late Rush for FCC Licenses," *Space News* (October 9–15, 1995):1; and Michael French, "Regulators to Choose Plan for Ka-Band Use," *Space News* (March 9–11, 1996):9; Conversation with Gedney.

Chapter 6 In the Wake of the *Challenger*
1. Jack Scarborough, "The Privatization of Expendable Launch Vehicles: Reconciliation of Conflicting Policy Objectives," *Policy Studies Review* 10(Spring–Summer 1991):23–24.

2. Ibid.; "Fletcher Nominated as New NASA Chief," *Science* 231 (March 21, 1986):1365. Beggs believed the indictment represented a White House attempt to move him out of NASA. He was subsequently completely exonerated. Joseph J. Trento, *Prescription for Disaster,* chap. 10.

3. Bruce A. Smith, "McDonnell Douglas Modifying Delta to Launch Navstar Spacecraft," *Aviation Week & Space Technology* 124(March 17, 1986):28. Conversation with Louis C. Raburn, 5 February 1995. House Committee on Science and Technology, "Assured Access to Space: 1986," Hearings, 99th Cong., 2d sess., February 26 -August 14, 1986, testimony on August 14, 1986, "Developing a U.S. ELV Capacity."

4. "A Mixed Fleet for NASA," *Science* 231(March 14, 1986):1238.

5. Gump, *Beyond NASA,* 91–92. David H. Moore, "Setting Space Transportation Policy for the 1990s: A Special Study," U.S. Congress, Congressional Budget Office, October 1986. Molly K. Macauley, "Space Transportation Policy: A Year of Upheaval," *Resources* (Winter 1987): 5–8.

6. Reed, *Implementation,* 127–28. The Department of Defense has often sought to encourage commercial counterparts to the military items it orders. (See, for example, Henry R. Hertzfeld, "The International Space Market," in *The U.S. Aerospace Industry: A Global Perspective for the 1990s* [Aerospace Industries Association, September 1991], and William L. Baldwin, *The Structure of the Defense Market, 1955–1964* [Durham, N.C.: Duke University Press, 1967], 176–77.)

7. Testimony of Richard Brackeen, in House Committee on Science and Technology, "Assured Access to Space: 1986," 592. See also M. Mitchell Waldrop, "Private Launch Prospects Improve," *Science* 236 (May 15, 1987): 766–68. Joan Johnson-Freese, *Changing Patterns of International Cooperation in Space* (Malabar, Fla.: Orbit Book Company, 1990), 63–65 and 73–74.

8. "Assured Access . . . Developing a Capacity," testimony of spokesmen for Martin Marietta, McDonnell Douglas, and General Dynamics. The Payload

Assist Module could be adapted for use with expendables, but it was more profitable to fly it on the shuttle (Conversation with John F. Yardley, November 15, 1995). Reed, *Implementation*, 262–87.

9. *Science* (1986):passim; Trento, *Prescription for Disaster*, chap. 10. House Committee on Science and Technology, "Assured Access to Space: 1986," testimony of William R. Graham, February 26, 1986.

10. *Weekly Compilation of Presidential Documents* 22, no. 33, 1103–4. Reed, *Implementation*, 54–57.

11. Scarborough, "Reconciliation," 22–23; House Committee on Science and Technology, "1987 NASA Authorization," February 25–April 10, 1986, 99th Cong., 2d sess., 472 and 527–54; "TCI Says Shuttle Price Should be $187 Million/ Asks $150 Million," *Defense Daily* (March 12, 1985):59.

12. *Defendant's Contention of Fact and Law in Transpace Carriers, Inc. vs. the United States,* NASA Historical Reference Collection, NASA Headquarters, Washington, D.C.

13. The United States, "Defendant's Contention," 27 *Federal Claims Reporter* 269(1992); Michael Isikoff, "NASA Ends Talks with Transpace," *Washington Post* (October 15, 1986):F1; "NASA Considers Ending Negotiations with Transpace Carriers over Rights to Delta Launcher," *Aviation Week & Space Technology* 125(October 13, 1986):26.

14. James C. Fletcher to David W. Grimes, October 10, 1986, and James C. Fletcher to Slade Gorton, November 3, 1986, File 010244, NASA Historical Reference Collection, NASA Headquarters, Washington, D.C.; Edmund L. Andrews, "Lost in Space," *Venture* 8(December 1986):38–42.

15. House Committee on Science and Technology, "Assured Access," 1986, 146–87. Conversation with George Baker, March 13, 1995; Theresa M. Foley, "Small Space Service Firms Do Well but Large Ventures Find Tough Going," *Aviation Week & Space Technology* 129(December 19, 1988):64–66. Astrotech Space Operations revived with the commercial launches of 1989 and went on to become a successful firm. Astrotech International left the space business and became a supplier of equipment for petroleum storage (telephone conversation with Don N. Stitt, February 25, 1996).

16. "OSC Offers to Finance Titan 34D for Mars Observer," *Aviation Week & Space Technology* 126(March 23, 1987):24–25; and "Orbital Sciences, Space Agency Disagree on Cost of Using Transfer Orbit Stage as Upper Stage for the Mars Observer," *Aviation Week & Space Technology* 126(May 11, 1987):27; "NASA Buys TOS for - What?" *Space Business News* (January 12, 1987):7–8. Orbital Sciences' offer was more complex than it is here represented, since NASA was to have the option of declining the Titan in October 1988.

17. House Committee on Science and Technology, "Assured Access," Hearings, 1986, 649 and 667. On AMROC's founding and its 1987 status, see David P. Gump, *Space Enterprise* (New York: Praeger, 1990), 31–35.

18. "Industry Swarms to Lightsats," *Space Business News* (August 10, 1987):2.

19. David W. Thompson, "The Microspace Revolution," address at Langley Colloquium Series, December 9, 1991 (Dulles, Va.: Orbital Sciences, 1992).

20. "OSC Targets July Test of Air-Launched Vehicle," *Space Business News* (May 30, 1988):3; and "Pegasus: Booster and Weapon All in One," *Space Business News* (June 13, 1988):1; Gump, *Space Enterprise*, 35–37.

21. "Lightsat Program Invests in Conestoga," *Space Business News* (January 25, 1988):1–2; "Army Funds Boost Private Rocket Firm," *Space Business News* (July 11, 1988):4–6, and "Black Monday Bludgeons Amroc," *Space Business News* (November 30, 1987):1–3.

22. Bruce A. Smith, "USAF Awards McDonnell Douglas Contract to Build, Operate MLVs," *Aviation Week & Space Technology* 126(January 26, 1987):20. The quotation is of Smith's paraphrase of Lt. Gen. Bernard Randolph.

23. Senate Committee on Armed Services, "Air Force Space Launch Policy and Plans," Hearings, October 6, 1987, 100th Cong., 1st sess., 15.

24. "Martin Teams with MD on MLV2," *Space Business News* (March 21, 1988):6–7.

25. Reed, *Implementation*, 126–29, quotation on 126.

26. Bruce D. Berkowitz, "Energizing the Space Launch Industry," *Issues in Science and Technology* 6, no. 2 (Winter 1989–90):77–83.

27. Berkowitz, "Energizing the Space Launch Industry."

28. Reed, *Implementation*, 200–201.

29. Library of Congress, Congressional Research Service, "Commercial Space Launch Services: The U.S. Competitive Position," Report for the House Committee on Science, Space, and Technology, November 1991, 22. The Navy and Strategic Defense Initiative office were far more willing to use launch services.

30. Reed, *Implementation*, 191–96. Congressional Office of Technology Assessment, *Access to Space: The Future of U.S. Space Transportation Systems*, (Washington D.C.: GPO, April 1990, OTA-ISC-415), 12–13.

31. Some of the regulatory problems are spelled out in testimony in Senate Committee on Commerce, Science, and Transportation, "Commercial Space Opportunities," Hearing, October 5, 1987, 100th Cong., 1st sess., 17–19.

32. Berkowitz, "Energizing the Space Launch Industry."

33. "Fawkes Looks to Make Space 'Just Another Place to Do Business,'" *Space Business News* (May 18, 1987):5; "New Policy Shifts Space Funding Emphasis," *Space Business News* (February 22, 1988):1.

34. U.S. Congress, Congressional Budget Office, *The NASA Program in the 1990s and Beyond* (Washington, D.C.: Congressional Budget Office, 1988), 80–82. CBO argued, "there is little clear evidence" for the first of these assumptions and less for the second.

35. "GOES Launcher Procurement Debated," *Aviation Week & Space Technology* 126(March 2, 1987):26; and "Commerce Dept. Will Buy ELVs From Private Sector" *Aviation Week & Space Technology* 126(March 23, 1987):24–25. "GD Gets First Government Launch Pact," *Space Business News* (November 2, 1987):6. "Commercial Space Launch Services," Library of Congress, Congressional Research Service, 16–21.

36. Theresa M. Foley, "Government Faulted for Frustrating Commercial Space Entrepreneurs," *Aviation Week & Space Technology* 128(February 15,

1988):79; and Theresa M. Foley, "NASA Wins Policy Dispute over Space Shuttle Pricing," *Aviation Week & Space Technology* (April 4, 1988):20.

37. Reed, *Implementation*, 61–66.

38. "Commercial Space Launch Services," Library of Congress Congressional Research Service. See also House Committee on Science, Space, and Technology, "Commercial Space Launch Act Implementation," Hearings, November 9, 1989, 101st Cong., 1st sess.

39. See Berkowitz, "Energizing the Space Launch Industry"; "Commercial Space Launch Services," Library of Congress Congressional Research Service; Congressional Office of Technology Assessment, *Access to Space*, chap. 6.

40. Senate Committee on Commerce, Science, and Transportation, "Commercial Space Opportunities," 54; Albert D. Wheelon, "The Future of the Unmanned Space Program," in *Space Policy Alternatives*, Radford Byerly, Jr., ed., (Boulder, Colo.: Westview Press, 1992), 13–34.

41. Brian G. Chow, "An Evolutionary Approach to Space Launch Commercialization," National Defense Research Institute, 1993, 30–31.

42. Macauley, "Space Transportation Policy."

43. Gruen, *Port Unknown,* 156–59; Lewin and Narayanan, *Keeping the Dream Alive.*

44. Gruen, *Port Unknown*, chap. 5, and Lewin and Narayanan, *Keeping the Dream Alive,* chap. 5 and 6.

45. Gruen, *Port Unknown,* and Levin and Narayanan, *Keeping the Dream Alive.*

46. House Committee on Science and Technology, "NASA's Space Station Activities," Hearings, August 2, 1983, 98th Cong., 1st sess., testimony of John Hodge and Peter Wood.

47. "Space-Lab Consortium in Works," and "Booz-Allen, Weinberg Report on Space Business Prospects," *Space Business News* (January 2, 1984):1 and 3.

48. House Committee on Science and Technology, "Space Commercialization," Hearings, May 3 ff, 1983, 98th Cong., 1st sess., 41.

49. Jay C. Lowndes, "Fairchild, NASA Agree on Leasecraft," *Aviation Week & Space Technology* 117(October 18, 1982):14–15; "Fairchild Seeks Agreements on Leasecraft," *Aviation Week & Space Technology* 120(June 25, 1984):54–55. "Fairchild Officials Townsend, Naugle Take Up Commercialization," *Space Business News* (January 30, 1984):4–5; "Development and Marketing Start on First Satellite for Hire," *Space World* (January 1984):10–11.

50. Eliot Marshall, "Space Stations in Lobbyland," and T. A. Heppenheimer, "Max and the Mini-Space Station," *Air & Space* (December 1988–January 1989):54–61; Conversation with Maxime A. Faget, January 30, 1995; Maxime A. Faget, testimony, House Committee on Science, Space and Technology, "1989 NASA Authorization," Hearings, March 23, 1988, 100th Cong., 2d sess., 444–85; Edwin E. Speaker to John Hodge, Robert Freitag, and Luther Powell, "ADB Meeting with Max Faget on December 15, 1983," File 010781, NASA Historical Reference Collection, NASA Headquarters, Washington, D.C.

51. Gump, *Space Enterprise*, 163–82; U.S. Department Of Commerce, *Space Commerce: An Industry Assessment* (Washington, D.C.: U.S. Department of

Commerce, May 1988), 79–96. Library of Congress Congressional Research Service, Tony Reichhardt, "U.S. Commercial Space Activities," February 1, 1992, 33–39.

52. "Spacehab Signs MOU with NASA," *Space Business News* (January 13, 1986):3; "Spacehab Courts Government Business," *Space Business News* (January 27, 1986); Department of Commerce, *Space Commerce*, 82–83.

53. Gruen, *Port Unknown*, chap. 5; Lewin and Narayanan, *Keeping the Dream Alive*, chap. 6.

54. Lewin and Narayanan, *Keeping the Dream Alive*, chap. 10, quotation on 99.

55. Gruen, *Port Unknown*, chap. 6 and 10.

56. Ibid., chap. 10.

57. "Grumman, Lockheed Win Station Deals," *Space Business News* (July 13, 1987):1–3, lists Grumman team duties as aid in engineering, integration, program management, information operations, utilization, safety, quality assurance, and international integration.

58. Conversation with Charles D. Walker, March 10, 1995. Charles D. Walker, "Pharmaceutical R & D in Space: An Industry Perspective," *Journal of Clinical Pharmacology* 31(1991):988–92.

59. "Ortho Div. Drops Initial Effort to Develop Medicine in Space," *Aviation Week & Space Technology* 123(September 16, 1985):21; Conversation with Charles D. Walker, March 10, 1995.

60. House Committee on Science, Space and Technology, "1989 NASA Authorization," Hearings, March 3 . . . , 1988, 100th Cong., 2d sess., testimony of James T. Rose, quotation on 590.

61. Craig Covault, "McDonnell Douglas, 3M Join to Produce Blood Drug in Space," *Aviation Week & Space Technology* 123 (November 18, 1985):1–17; Bruce A. Smith, "McDonnell Douglas Plans to Process Large Pharmaceutical Batch in Space," *Aviation Week & Space Technology* 123(November 18, 1985): 83–85; "McDonnell Douglas, 3M Halt Hormone Production Plans," *Aviation Week & Space Technology* 124(March 17, 1986):32; "Entrepreneurs Undaunted," *Space Business News* (January 30, 1986):1ff.

62. "Shuttle Uncertainties Tied to Space Commercialization Risks," *Aerospace Daily* (January 17, 1985):90.

63. Theresa M. Foley, "NASA Denies Shuttle Launch to Commercial Space Firm," *Aviation Week & Space Technology* 127(September 7, 1987):26–28; Foley, "The Broken Promise of Commercial Space," *Aviation Week & Space Technology* (September 14, 1987):15; Conversations with Dale D. Myers, January 17, 1994, and Isaac T. Gillam, February 20, 1996.

64. Theresa M. Foley, "Microgravity Task Force Recommends Sweeping Changes to NASA Program," and "NASA Issues Five Directives on Commercial Space Policy," *Aviation Week & Space Technology* 127(July 20, 1989):50–54.

65. Craig Covault, "Lack of Insurance, Customers, Halts Fairchild Leasecraft," *Aviation Week & Space Technology* 123(November 11, 1985):16–17.

66. House Committee on Science, Space and Technology, "1989 NASA Authorization," quotation on 479. Marshall, "Lobbyland."

67. Marshall, "Lobbyland." Gruen, *Port Unknown,*" chap. 10 and epilogue.

68. Craig Covault, "NASA Preparing 412 Space Station Contracts," *Aviation Week & Space Technology* 121(September 17, 1984):16–17; Angelo Guastaferro, "The Role of Incentives and Accountability in Industry and Government," in Radford Byerly, Jr., ed., *Space Policy Alternatives,* (Boulder, Colo.: Westview Press, 1992), chap. 7.

69. Guastaferro, "Role of Incentives," 113. Jacques Gansler, *The Defense Industry,* (Cambridge, Mass.: MIT Press, 1980), chap. 2.

70. Dwayne A. Day, "Doomed to Fail: The Birth and Death of the Space Exploration Initiative," *Spaceflight* 37, no. 3(March 1995):79–83.

71. Theresa M. Foley, "Six Aerospace Firms Compete to Build Space Station Hardware," *Aviation Week & Space Technology* 127(July 27, 1987):26–27.

72. Aerospace Industries Association, *Aerospace Facts and Figures, 1989–1990* (Washington, D.C.: Aerospace Industries Association, 1990); Richard G. O'Lone, "Mesa Purchase of Boeing Stock Highlights Industry Pressures," *Aviation Week & Space Technology* 127(August 3, 1987):24–25.

73. This spurt ended as the 1990s began. See Marie-Christine Schmitt, "Aéronautique: Une crise mondiale, des industries en mutation," *Problèmes économiques* no. 2, 405(January 4, 1995):25–31. For other treatments of the contraction in industry markets, see Keith Hayward, *The World Aerospace Industry: Collaboration and Competition* (London: Duckworth and RUSI, 1994), and John A. Alic et al., *Beyond Spinoff: Military and Commercial Technologies in a Changing World* (Boston: Harvard Business Press, 1992).

74. AIA, *Facts & Figures, 1989–1990.*

75. In 1993 the head of the Aerospace Industries Association estimated the share of commercial space as 14 percent. House Committee on Science, Space and Technology, "Future of the U.S. Space Industrial Base," Hearings, February 2 . . . , 1993, 103d Cong., 1st sess., testimony of Don Fuqua.

76. Steven Perlstein, "Space-Defense's Final Frontier or Last Hope?" *Washington Post* (June 6, 1991):C10.

77. Gruen, *Port Unknown,* 110–17.

78. Ibid., chap. 10 and epilogue.

79. In addition to the previous references, see the testimonies of Gregg R. Fawkes and Raymond G. Kemmer in House Committee on Science, Space and Technology, "1989 NASA Authorization." Theresa M. Foley, "Government Faulted for Frustrating Commercial Space Entrepreneurs," *Aviation Week & Space Technology* 128(February 15, 1988):79.

80. House Committee on Science, Space, and Technology, "1989 NASA Authorization," 482. The remark was made, appropriately enough, by Joseph P. Allen, IV, then Space Industries executive vice president, but formerly an astronaut.

81. Ibid., Hearings, see the testimony of Andrew J. Stofan, associate administrator for the Space Station.

82. Ibid., 412.

83. Ibid., 679 and 678.

84. Ibid., 646. By "suboptimal," Beggs meant that, because it had already

committed to the rental, NASA would be tempted to use the CDSF even in cases where the Spacehab would be more appropriate.

85. "'NASA Abandons CDSF: Supports Private Sector Involvement,'" *Aerospace Daily* (May 17, 1989). "Commercial Space Facility in for Rough Ride," *Space Business News* (April 4, 1988):2.

86. Liz Tucci, "Report Says NASA should Drop Spacehab," *Space News* (October 11–17, 1993):1. Spacehab and the ISF were generally judged the worthiest of these projects. Etco's Labitat had the drawback that it is orders of magnitude more expensive to retrofit a fuel tank into a laboratory in space than to fit up a laboratory on the ground.

Chapter 7 Trends in NASA-Industry Relations

1. Raymond L. Garthoff, *The Great Transition: American-Soviet Relations and the End of the Cold War* (Washington, D.C.: Brookings Institution, 1994).

2. U.S. House, Committee on Science, "NASA Procurement in the Earth-Space Economy," Nov. 8, 1995, 104th Cong., 1st sess., 2.

3. *Aeronautics and Space Report of the President: Fiscal Year 1995 Activities* (Washington D.C.: NASA), A-31. Budgets and budget deficits come from the 1996 edition of *Statistical Abstracts of the U.S.*

4. Jacques S. Gansler, *Defense Conversion: Transforming the Arsenal of Democracy* (Cambridge, Mass.: MIT Press, 1995); Aerospace Industries Association, *Aerospace Facts & Figures, 1994–1995* (Washington, D.C.: Aerospace Industries Association of America, 1994).

5. Congress Joint Economic Committee, "The Aerospace Industry," hearings, 102d Cong., 1st and 2d sess., December 3, 1991 and February 27, 1992.

6. Anne Eisele, "Boeing Declines to Bid on $6 Billion Contract," *Space News* 8, no. 11(March 17–23, 1997):3.

7. House Committee on Science, Space, and Technology, "The Future of the U.S. Space Industrial Base," Hearings, 103d Cong., 1st sess., February 1993, 73; Aerospace Industries Association, "1993 Year-End Review and Forecast" (Available from AIA, 1250 Eye Street NW, Washington, D.C.), table 1.

8. Warren S. Watkins, "Single Purpose Satellite Systems," in *Proceedings of the Third Annual AIAA/Utah State University Conference on Small Satellites* (Logan: Utah State University, 1989).

9. John Carey, "The Next Space Race: Snapshots?," *Business Week* (December 11, 1995):111–12. It is worth remarking that more than one of these are cross-national ventures.

10. Michael Harr and Rajiv Kohli, *Commercial Utilization of Space: An International Comparison of Framework Conditions* (Columbus, Ohio: Battelle Press, 1990), 58.

11. Many of the small companies started before 1990 are described in Gump, *Space Enterprise.* For the firms mentioned here, see Charles Petit, "Economy Rocket Get Tow Part Way," *San Francisco Chronicle* (October 30, 1996); Warren Ferster, "Loral Signs on With Kistler," *Space News* 8, no. 5(February 3–9, 1997): 6. Small companies appear to be increasingly carrying out their projects by means of joint ventures with larger firms.

12. "Truly's Dismissal Puts NASA on Autopilot," *Science* 255(February 21, 1992):915; "NASA Mystery Man," *Science* 255(March 20, 1992):1506; William J. Broad, "NASA Chief Quits in Policy Conflict," *New York Times* (February 13, 1992):A, 1:1; and Warren E. Leary, "Quayle's Influence Seen in NASA Shake-Up," *New York Times* (February 15, 1992):I, 7:4.

13. Dwayne A. Day, "Doomed to Fail: The Birth and Death of the Space Exploration Initiative," *Spaceflight* 37, no. 3(March 1995):79–83.

14. U.S. Senate, Committee on Commerce, Science, and Transportation, Hearing, "NASA's Fiscal Year 1994 Budget," 103/1st, April 20, 1993; U.S. House Committee on Science, Space and Technology, "National Space Transportation Policy," Hearing, 103d Cong., 2d sess., September 20, 1994. David C. Morrison, "Low-Rent Space," *National Journal* 29(1995):1028–32.

15. House Committee on Science, Space, and Technology, 101st Cong., 2d sess. "1991 NASA Authorization," Vol. 2, Hearings, February 6, 1990 . . . , 157–58. Conversation with Willis M. Hawkins, September 7, 1994.

16. Conversation with John B. Winch, November 14, 1994. Andrew Lawler, "Station Plan Stuns Skeptics," *Space News* 5, no. 14(April 4–10, 1994):1 and 28.

17. William Harwood, "Shuttle Budget at $3.1 Billion," *Space News* 5, no. 39(October 10–16, 1994):4; William Harwood, "NASA Panel to Review Shuttle Program," *Space News* 5, no. 46(December 5–11, 1994):4; Ben Iannotta, "NASA's Future Lean," *Space News* 6, no. 5(February 6–12, 1995):4; William Harwood, "Thousands of Jobs at Stake," *Space News* 6, no. 9(March 6–12, 1995):3; William Harwood, "USA Chief Says Shuttle May Linger Until 2020," *Space News* 7, no. 17(April 29–May 5, 1996):3.

18. U.S. Congressional Office of Technology Assessment, *The National Space Transportation Policy: Issues for Congress*, OTA-ISS-620, May 1995; Senate Committee on Commerce, Science, and Transportation, "The NASA Space Shuttle and the Reusable Launch Vehicle Program," Hearing, 104th Cong., 1st sess., May 16, 1995; House Committee on Science, "The X-33 Reusable Launch Vehicle: A New Way of Doing Business?" Hearing, 104th Cong., 1st sess., November 1, 1995; *Space News* (1994, 1995, 1996):passim.

19. "NASA Implementation Plan for the National Space Transportation Policy" (November 7, 1994, revised, November 21, 1995). I am indebted to Andrew J. Butrica, X-33 project historian, for this and other materials and information.

20. Senate Committee on Commerce, Science, and Transportation, "NASA Space Shuttle and the Reusable Launch Vehicle," 13.

21. Ben Iannotta, "OSC, Rockwell Selected to Run X-34 Project," *Space News* 6, no. 10(March 13–19, 1995):4; and William Boyer, "X-33 Designs Touted as Deadline Nears," *Space News* 7, no. 12(March 25–31, 1996):4.

22. Ben Iannotta, "White House Revives X-34 Partnership," *Space News* 6, no. 43(November 6–12, 1995):1; Ben Iannotta, "NASA, Industry Strike X-34 Deal," *Space News* 6, no. 45(November 27–December 3, 1995):1; Ben Iannotta, "X-34 Partners Review Design, Suspend Work," *Space News* 7, no. 5(February 5–11, 1996):1; Ben Iannotta, "NASA Working to Minimize X-34 Fallout," *Space News* 7, no. 7(February 19–25, 1996):4; Anne Eisele, "Orbital Sciences Gets X-34 Nod Again," *Space News* 7, no. 24(June 17–23, 1996):4.

23. Office of Technology Assessment, *National Space Transportation Policy*, gives this misgiving, and some of industry's others. See also Don Carney, "X-33 Men," *Washington Monthly* 28, No. 12 (December 1996): 24–27, and Peter Spiegel, "Free Launch?" *Forbes* 159, No. 4 (February 24, 1997): 76.

24. House Committee on Science, "The X-33," testimonies of Robert G. Minor, president, Space Systems Division of Rockwell International, 31, and NASA Associate Administrator John E. Mansfield, 98.

25. David L. Chandler, "NASA Quest to Cut Space Travel Cost," *Boston Globe*, reprinted in the *San Francisco Chronicle* (June 20, 1997):A10; Anne Eisele, "NASA will Test Small Boosters," *Space News* 8, no. 24(June 16–22, 1997):4.

26. Warren Ferster, "NASA Seeks EOS Cost Cuts: Production-Line Satellites Seen as Likely Option," *Space News* 7, no. 25(June 24–30, 1996):1. Ferster, "Industry Advice Leads to Change in Light SAR," *Space News* 7, no. 34(September 2–8, 1996):1.

27. McCurdy, *Inside NASA*, 28–39, 93–96.

28. Report of the Advisory Committee on the Future of the U.S. Space Program, December 1990, 41–42.

29. Leonard David, "Moon Mission Has Hidden Goal: Project is a Showcase for Low-Cost, Quick Methods," *Space News* 7, no. 14(April 8–14, 1996):31.

30. Anne Eisele, "NASA Adds More Duties to Privatization Proposal," *Space News* 8, no. 8(February 24–March 2, 1997):3, 19; Warren Ferster, "NASA will Award Projects to Test Data Systems," *Space News* 7, no. 28(July 15–21, 1996):6.

31. On national industrial cultures, see Kenneth Dyson, "The Cultural, Ideological, and Structural Context," in *Industrial Crises: A Comparative Study of the State and Industry*, Kenneth Dyson and Stephen Wilks, eds. (Oxford: Martin Robertson, 1983), 26–66.

32. For this view of NASA, see Gump, *Space Enterprise*.

33. One interesting perspective on this change is the article by Patrick J. Akard, "Corporate Mobilization and Political Power: The Transformation of US Economic Policy in the 1970s," *American Sociological Review* 57(October 1992):597–615.

34. Marcia S. Smith, "A Congressional Perspective on the Space Exploration Initiative," in *Mars: Past, Present, and Future*, E. Brian Pritchard, ed., Progress in Astronautics and Aeronautics, Vol. 145 (Washington, D.C.: AIAA, 1993).

35. Roger Handberg, *The Future of the Space Industry: Private Enterprise and Public Policy* (Westport, Conn.: Quorum Books, 1995).

Bibliography

This book is based mainly on published sources of three kinds: secondary books and articles, trade and business journals, and U.S. government publications, including congressional hearings and reports by the congressional Office of Technology Assessment. Among trade publications, I made especial use of *Aviation Week & Space Technology*. It is a goldmine of information on space history, and is studded with quotations and paraphrases that give insight into NASA and industry preoccupations and viewpoints. For the middle 1980s, I was able to use *Space Business News* through the kind offices of David P. Gump. For the 1990s, I relied heavily on *Space News*.

The book makes relatively modest use of manuscripts. I began my research with roughly forty informal conversations with engineers and managers from NASA and industry. The conversations were not taped but I have vetted those pages that rely upon them with my informants.

The principal collection consulted was the NASA Historical Reference Collection at NASA Headquarters in Washington, D.C. This is a group of heterogeneous materials—NASA documents, newspaper and magazine clippings, unpublished manuscripts commissioned by the History Office, and more—selected over many years for their historical importance by NASA archivists and historians. I spent a few weeks at the excellent archive of director's files Marshall Space Flight Center historian Michael Wright has assembled. I also made use of the fine interviews on the Apollo and shuttle projects at the Johnson Space Center's History Office, and the interviews on space in the National Air and Space Museum's Glennan-Webb-Seamans Interview Project.

Rockwell International Corporation gave me the opportunity to examine its company archive in Seal Beach, California, which is weighted toward photographs, publicity releases, and internal publications. Hughes Aircraft Company records managers went out of their way to provide access to Hughes' Syncom files. I found, however, that this very good collection had already been well mined by other authors. Finally, many of those whom I interviewed graciously provided me with copies of documents from their personal files.

I looked for sources that would enable me to get a handle on the book's topics. I did not have time to study all the available literature on each subject, nor did I go through NASA or industry archives in the systematic fashion that I, as a historian, would have preferred. The book that resulted, therefore, necessarily fits somewhere between the genres of popular and scholarly literature.

A principal purpose of this bibliography is to make it easy for readers to find full citations to works mentioned in the notes. Secondary works and government publications are listed separately, the former alphabetically by author and the latter by agency or branch, and then by date. The trade and business journal articles cited in the notes are not included here. Also omitted are a few secondary works not concerned with NASA or the aerospace industry nor cited more than once.

SECONDARY WORKS

Alic, John A., Harvey Brooks, Lewis H. Branscomb, Ashton B. Carter, and Gerald L. Epstein. *Beyond Spinoff: Military and Commercial Technologies in a Changing World*. Boston: Harvard Business Press, 1992.

Armacost, Michael H. *The Politics of Weapons Innovation: The Thor-Jupiter Controversy*. New York: Columbia University Press, 1969.

Armstrong, Philip, Andrew Glyn, and John Harrison. *Capitalism since 1945*. Oxford: Basil Blackwell, 1991.

Baldwin, William L. *The Structure of the Defense Market, 1955–1964*. Durham N.C.: Duke University Press, 1967.

Banks, Jeffrey S. "The Space Shuttle." In *The Technology Pork Barrel*, edited by Linda R. Cohen and Roger G. Noll. Washington, D.C.: The Brookings Institution, 1991. 179–215.

Barfield, Claude E. "Technology Report/NASA Broadens Defense of Space Shuttle to Counter Critics' Attacks." *National Journal* 4(19 August 1972):1323–32.

Beard, Edmund. *Developing the ICBM: A Study in Bureaucratic Politics*. New York: Columbia University Press, 1976.

Berkowitz, Bruce D. "Energizing the Space Launch Industry." *Issues in Science and Technology* 6, no. 2(Winter 1989–90):77–83.

Bilstein, Roger E. *Stages to Saturn: A Technological History of the Apollo/Saturn Launch Vehicle*. Washington D.C.: NASA Scientific and Technical Information Branch, 1980.

———. *Flight in America, 1900–1983: From the Wrights to the Astronauts*. Baltimore: Johns Hopkins University Press, 1984.

———. *The American Aerospace Industry: From Workshop to Global Enterprise*. New York: Twayne, 1996.

Bright, Charles D. *The Jet Makers: The Aerospace Industry from 1945 to 1972.* Lawrence: Regents Press of Kansas, 1978.

Brooks, Courtney G., James M. Grimwood, and Loyd S. Swenson, Jr. *Chariots for Apollo: A History of Manned Lunar Spacecraft.* Washington, D.C.: NASA, 1979.

Bugos, Glenn E. "Program Management in the Department of Defense and NASA: . . . ," HHN 157, NASA Historical Reference Collection, NASA Headquarters, Washington, D.C., March 15, 1987, photocopy.

———. "Manufacturing Certainty: Testing and Program Management for the F-4 Phantom II." *Social Studies of Science* 23(1993):265–300.

Burrows, William E. "The Military in Space: Securing the High Ground." In *Space: Discovery and Exploration,* edited by Martin J. Collins and Sylvia K. Kraemer, 116–65. Hong Kong: Hugh Lauter Levin, 1993.

Cargill-Hall, R. "The Eisenhower Administration and the Cold War: Framing American Astronautics to Serve National Security." In *Organizing for the Use of Space: Historical Perspectives on a Persistent Issue,* edited by Roger D. Launius, 49–61. AAS History Series, Vol. 18. San Diego: Univelt, 1995.

Chang, Ha-Joon. *The Political Economy of Industrial Policy.* New York: St. Martin's, 1994.

Chow, Brian G. "An Evolutionary Approach to Space Launch Commercialization." National Defense Research Institute, 1993.

Cohen, Linda R., and Roger G. Noll, eds. *The Technology Pork Barrel.* Washington, D.C.: Brookings Institution, 1991.

Collins, Guy. *Europe In Space.* London: Macmillan, 1990.

Collins, Robert M. "Growth Liberalism in the Sixties: Great Societies at Home and Grand Designs Abroad." In *The Sixties . . . From Memory to History,* edited by David Farber, 11–44. Chapel Hill: University of North Carolina Press, 1994.

Corliss, William R. *History of the Delta Launch Vehicle.* NASA History Office, File 010246, "Delta Documentation (1959–1972)," NASA Historical Reference Collection, Washington, D.C., June 1972, photocopy.

Cunniffe, Peter. *Misreading History: Government Intervention in the Development of Commercial Communications Satellites,* Program in Science and Technology for International Security Report No. 24. Cambridge, Mass.: MIT, May 1991.

Danhof, Clarence A. *Government Contracting and Technological Change.* Washington, D.C.: Brookings Institution, 1968.

Dawson, Virginia P. *Engines and Innovation: Lewis Laboratory and American Propulsion Technology.* Washington D.C.: NASA, 1991.

———. "Building Space Hardware: Industry, National Security, and NASA." In *Space: Discovery and Exploration,* edited by Martin J. Collins and Sylvia K. Kraemer, 166–219. Washington, D.C.: Hugh Lauter Levin for Smithsonian Institution, 1993.

Day, Dwayne A. "Doomed to Fail: The Birth and Death of the Space Exploration Initiative." *Spaceflight* 37, no. 3(March 1995): 79–83.

———. "A Strategy for Space." *Spaceflight* 38(September 1996):308–12.

————. "Invitation to Struggle: The History of Civilian-Military Relations in Space." In *Exploring the Unknown*, Vol. 2, *External Relationships*, edited by John M. Logsdon, 262–63. Washington, D.C.: NASA, 1996.

Dethloff, Henry C. *Suddenly Tomorrow Came . . . A History of the Johnson Space Center*. Houston: NASA Lyndon B. Johnson Space Center, 1993.

DeVorkin, David. *Science With a Vengeance: The Military Origins of Space Sciences in the American V-2 Era*. New York: Springer, 1992.

Divine, Robert A. "Lyndon B. Johnson and the Politics of Space." In *The Johnson Years*, Vol. 2, *Vietnam, the Environment, and Science*, edited by Robert A. Divine, 217–53. Lawrence: University Press of Kansas, 1987.

————. *The Sputnik Challenge*. New York: Oxford University Press, 1993.

Dupré, J. S., and W. E. Gustafson. "Contracting for Defense: Private Firms and the Public Interest." *Political Science Quarterly* 77(1962):161–77.

Dyer, David, and David B. Sicilia. *Labors of a Modern Hercules: The Evolution of a Chemical Company*. Boston: Harvard Business School, 1990.

Elder, Donald C. *Out from Behind the Eight Ball: A History of Project Echo*. American Astronautical Society History Series, Vol. 16. San Diego: Univelt, 1995.

Fries, Sylvia D. "2001 to 1994: Political Environment and the Design of NASA's Space Station System." *Technology and Culture* 29(1988):568–93.

————. *NASA Engineers and The Age of Apollo*. Washington, D.C.: NASA, 1991.

Galloway, Jonathan F. *The Politics and Technology of Satellite Communications*. Lexington, Mass.: Lexington Books, 1972.

Gansler, Jacques S. *The Defense Industry*. Cambridge, Mass.: MIT Press, 1982.

————. *Defense Conversion: Transforming the Arsenal of Democracy*. Cambridge, Mass.: MIT Press, 1995.

Glennan, T. Keith. *The Birth of NASA: The Diary of T. Keith Glennan*, edited by J. D. Hunley. Washington, D.C.: NASA, 1993.

Glover, Daniel R. "NASA Experimental Communications Satellites, 1958–1995." In *Beyond the Ionosphere: Fifty Years of Satellite Communication*, edited by Andrew J. Butrica, 51–64. Washington, D.C.: NASA History Office, 1997.

Goldman, Nathan C. *Space Commerce: Free Enterprise on the High Frontier*. Cambridge, Mass.: Ballinger, 1985.

Gray, Mike. *Angle of Attack: Harrison Storms and the Race to the Moon*. New York: W. W. Norton, 1992.

Green, Constance McLaughlin, and Milton Lomask. *Vanguard, A History*. Washington, D.C.: Smithsonian Institution Press, 1971.

Griffith, Allison. *The National Aeronautics and Space Act: A Study of the Development of Public Policy*. Washington, D.C.: Public Affairs Press, 1962.

Gruen, Adam L. *The Port Unknown: A History of the Space Station Freedom Program*. NASA Historical Reference Collection, Washington, D.C., April 1993, photocopy.

Guastaferro, Angelo. "The Role of Incentives and Accountability in Industry and Government." In *Space Policy Alternatives*, edited by Radford Byerly, Jr., chap. 7. Boulder, Colo.: Westview Press, 1992.

Guilmartin, John F., Jr., and John Walker Mauer. *A Shuttle Chronology, 1964–1973*, JSC 23309. Houston: NASA, Johnson Space Center, 1988.

Gump, David P. *Space Enterprise: Beyond NASA*. New York: Praeger, 1990.

Hacker, Barton C., and James M. Grimwood. *On the Shoulders of Titans: A History of Project Gemini*. Washington, D.C.: NASA Scientific and Technical Information Office, 1977.

Haley, Andrew G. *Rocketry and Space Exploration*. Princeton: Van Nostrand, 1958.

Hallion, Richard. *Supersonic Flights: The Story of the Bell X-1 and Douglas D-558*. New York: Macmillan, 1972.

Handberg, Roger. *The Future of the Space Industry: Private Enterprise and Public Policy*. Westport, Conn.: Quorum Books, 1995.

Hansen, James R. *Spaceflight Revolution: NASA Langley Research Center From Sputnik to Apollo*. Washington, D.C.: NASA, 1995.

Harr, Michael, and Rajiv Kohli, *Commercial Utilization of Space: An International Comparison of Framework Conditions*. Columbus, Ohio: Battelle Press, 1990.

Harwood, William B. *Raise Heaven and Earth: The Story of Martin Marietta People and their Pioneering Achievements*. New York: Simon & Schuster, 1993.

Hayward, Keith. *The World Aerospace Industry: Collaboration and Competition*. London: Duckworth and Royal United Services Institute for Defence Studies (RUSI), 1944.

Hechler, Ken. *The Endless Space Frontier: A History of the House Committee on Science and Astronautics, 1959–1978*, abridged by Albert E. Eastman. American Astronautical Society History Series, Vol. 4. San Diego: Univelt, 1982.

Heiss, Klaus P. "A Perspective on Private Initiatives in U.S. Space Transportation." *Journal of Social, Political, and Economic Studies* 11, no. 1(Spring 1986): 3–16.

Homann, Robert F.S. "Weapons System Concepts and Their Pattern in Procurement." *Federal Bar Journal* 17(1957):402–19.

Hunley, J.D. "The Evolution of Solid-Propellant Rocketry in the United States." *Quest: The History of Spaceflight Quarterly* 6, no. 1(1998):22–38.

Hutchins, Dexter C. "Entrepreneurs Aim for Outer Space." *Venture* 2(September 1980):48–50.

Jaffe, Leonard. *Communication in Space*. New York: Holt, Rinehart and Winston, 1966.

Johnson-Freese, Joan. *Changing Patterns of International Cooperation in Space*. Malabar Fla.: Orbit Book Company, 1990.

Kemp, Tom. *The Climax of Capitalism: The U.S. Economy in the Twentieth Century*. London: Longman, 1990.

Kloman, Erasmus H. *NASA: The Vision and the Reality*. Washington, D.C.: National Academy of Public Administration, 1985.

Koppes, Clayton R. *JPL and the American Space Program*. New Haven: Yale University Press, 1982.

Kovalick, Walter W., Jr. "Space Launch Services: Public or Private Operation?"

File 010782, NASA Historical Reference Collection, Washington, D.C., May 1982.

Kvam, Roger A. "Comsat: The Inevitable Anomaly." In *Knowledge and Power: Essays On Science and Government*, edited by Sanford A. Lakoff, 271–92. New York: Free Press, 1966.

Lambright, W. Henry. *Shooting Down the Nuclear Plane*. Inter-University Case Program No.104. Indianapolis: Bobbs-Merrill, 1967.

———. *Powering Apollo: James E. Webb of NASA*. Baltimore: Johns Hopkins University Press, 1995.

Launius, Roger D. "Toward an Understanding of the Space Shuttle: A Historiographical Essay." *Air Power History* 39(Winter 1992):3–18.

———. Introduction to *The Birth of NASA: The Diary of T. Keith Glennan*, edited by J.D. Hunley. Washington D.C.: NASA, 1993.

———. *NASA: A History of the U.S. Civil Space Program*. Malabar, Fla.: Krieger, 1994.

———. "Early U.S. Civil Space Policy, NASA, and the Aspiration of Space Exploration." In *Organizing for the Use of Space: Historical Perspectives on a Persistent Issue*, edited by R. D. Launius, 63–86. AAS History Series, Vol. 18. San Diego: Univelt, 1995.

———. "A Waning of Technocratic Faith: NASA and the Politics of the Space Shuttle Decision." *Journal of the British Interplanetary Society* 49(1996): 49–58.

Launius, Roger D., and Aaron K. Gillette, *Toward A History of the Space Shuttle: An Annotated Bibliography*, Studies in Aerospace History no.1. Washington D.C.: NASA History Office, December 1992.

Launius, Roger D., and J. D. Hunley. *An Annotated Bibliography of the Apollo Program*. NASA History Office Monographs in Aerospace History No. 2, Washington, D.C.: NASA, July 1994.

Levine, Arnold S. *Managing NASA in the Apollo Era*. Washington, D.C.: NASA, 1982.

Lewin, Thomas J., and V. K. Narayanan. *Keeping the Dream Alive: Managing the Space Station Program, 1982–1986*. NASA Contractor Report No. 4272. Washington, D.C.: NASA, July 1990.

Lindley, Daniel. "No Free Launch: Tracking Space Fledglings." *Barron's* 64(September 17, 1984):47ff.

Livingston, J. Sterling. "Weapon System Contracting." *Harvard Business Review* 37(July-August 1959):83–92.

Logsdon, John M. "The Space Shuttle Decision: Technology and Political Choice." *Journal of Contemporary Business* 7(1978):13–29.

———. "From Apollo to Shuttle: Policy Making in the Post-Apollo Era." NASA Historical Reference Collection, Washington, D.C., Spring 1983, photocopy.

———. "The Space Shuttle Program: A Policy Failure?" *Science* 232(30 May 1986):1099–1105.

———. "Choosing Big Technologies: Examples from the U.S. Space Program." *History and Technology* 9(1992):139–50.

Logsdon, John M., ed. *Exploring the Unknown: Selected Documents in the His-*

tory of the U.S. Civil Space Program, Vol. 1, Organizing for Exploration. Washington D.C.: NASA, 1995. Vol. 2, External Relationships. Washington, D.C.: NASA, 1996.

Macauley, Molly K. "Space Transportation Policy: A Year of Upheaval." Resources (Winter 1987):5–8.

———. "Rethinking Space Policy: The Need to Unearth the Economics of Space." In Space Policy Reconsidered, edited by Radford Byerly, Jr., 131–43. Boulder, Colo.: Westview Press, 1989.

McCurdy, Howard E. The Space Station Decision: Incremental Politics and Technological Choice. Baltimore: Johns Hopkins University Press, 1990.

———. Inside NASA: High Technology and Organizational Change in the U.S. Space Program. Baltimore: Johns Hopkins University Press, 1993.

McDougall, Walter A. . . . The Heavens and The Earth: A Political History of The Space Age. New York: Basic Books, 1985.

McKay, W. D. Can Democracies Fly in Space?: The Challenge of Revitalizing the U.S. Space Program. Westport, Conn.: Praeger, 1995.

McQuaid, Kim. Uneasy Partners: Big Business in American Politics, 1945–1990. Baltimore: Johns Hopkins University Press, 1994.

Mack, Pamela E. Viewing the Earth: The Social Construction of the Landsat Satellite System. Cambridge, Mass.: MIT Press, 1990.

Marshall, Eliot. "Space Stations in Lobbyland." In Air & Space (December 1988–January 1989):54–61.

Morse, Edgar W. "Preliminary History of the Origins of Project Syncom." NASA Historical Note No. 44, NASA Historical Reference Collection, Washington, D.C., photocopy.

Mowery, David C. Alliance Politics and Economics: Multinational Joint Ventures in Commercial Aircraft. Cambridge, Mass.: Ballinger, 1987.

Murphy, Thomas P. "The Manned Orbiting Laboratory Controversy." In Science, Geopolitics, and Federal Spending, Thomas P., Murphy, 463–87. Lexington, Mass.: Heath Lexington Books (1971).

Murray, Charles, and Catherine Bly Cox. Apollo: The Race to the Moon. New York: Simon & Schuster, 1989.

Musolf, Lloyd D., ed. Communications Satellites in Political Orbit. San Francisco: Chandler, 1968.

Neufeld, Jacob. The Development of Ballistic Missiles in the United States Air Force, 1945–1960. Washington, D.C.: Office of Air Force History of the U.S. Air Force, 1990.

Pearcy, Arthur. Flying the Frontiers: NACA and NASA Experimental Aircraft. Annapolis: Naval Institute Press, 1993.

Perry, Nancy O. "Shooting for the Stars." In Harvard Business School Bulletin (June 1989):2–11.

Pielke, Roger A., Jr., and Radford Byerly, Jr. "The Space Shuttle Program: Performance Versus Promise." In Space Policy Alternatives, edited by Radford Byerly, Jr., 223–45. Boulder, Colo.: Westview Press (1992).

Pierce, J. R. The Beginnings of Satellite Communications. San Francisco: San Francisco Press, 1968.

Rae, John B. *Climb to Greatness: The American Aircraft Industry, 1920–1960*. Cambridge, Mass.: MIT Press, 1968.

Reed, Craig Russell. "U.S. Commercial Space Launch Policy Implementation, 1986–1992." Ph.D. dissertation, George Washington University, Washington, D.C., 1998.

Reichart, Otto H. "Industrial Concentration and World War II: A Note on the Aircraft Industry." *Business History Review* 49(1975):498–503.

Reiss, Edward. *The Strategic Defense Initiative*. Cambridge: Cambridge University Press, 1992.

Richelson, Jeffrey T. *America's Secret Eyes in Space: The U.S. Keyhole Spy Satellite Program*. New York: Harper & Row, 1990.

Roland, Alex. *Model Research: The National Advisory Committee for Aeronautics, 1915–1958*. Washington D.C.: NASA Scientific and Technical Branch, 1985.

———. "The Shuttle: Triumph or Turkey?" *Discover* 6(November 1985):29–49.

Scarborough, Jack, "The Privatization of Expendable Launch Vehicles: Reconciliation of Conflicting Policy Objectives." *Policy Studies Review* 10 (Spring/Summer 1991):12–30.

Schmitt, Marie-Christine. "Aéronautique: Une crise mondiale, des industries en mutation." In *Problèmes économiques* 405, no. 2(January 4, 1995):25–31.

Schoettle, Enid Curtis Bok. "The Establishment of NASA." In *Knowledge and Power: Essays on Science and Government*, edited by Sanford A. Lakoff, 162–270. New York: Free Press, 1966.

Seamans, Robert C., Jr. *Aiming at Targets*. Washington, D.C.: NASA, 1996.

Simonson, G. R., ed. *The History of the American Aircraft Industry*. Cambridge, Mass.: MIT Press, 1968.

Smith, Marcia S. "Civilian Space Applications: The Privatization Battleground." In *Space Policy Reconsidered*, edited by Radford Byerly, Jr., 105–16. Boulder, Colo.: Westview Press, 1989.

———. "A Congressional Perspective on the Space Exploration Initiative." In *Mars: Past, Present, and Future*, edited by Brian Pritchard, 109–13. Progress in Astronautics and Aeronautics, Vol. 145. Washington, D.C.: American Institute of Aeronautics and Astronautics (AIAA), 1993.

Stares, Paul B. *The Militarization of Space: U.S. Policy, 1945–1984*. Ithaca, N.Y.: Cornell University Press, 1985.

Steinberg, Gerald M. "The Militarization of Space: From Passive Support to Actual Weapons Systems." In *The Exploitation of Space: Policy Trends in the Military and Commercial Uses of Outer Space*, edited by Michiel Schwarz and Paul Stares, 31–49. London: Butterworth, 1985.

Stekler, Herman O. *The Structure and Performance of the Aerospace Industry*. Berkeley: University of California Press, 1965.

Stevenson, Adlai E. "The New Era in Space." *Journal of* Contemporary *Business* 7, no. 3(1978):7–12.

Stuhlinger, Ernst, and Frederick I. Ordway, III. *Wernher von Braun, Crusader for Space: A Biographical Memoir*. Malabar, Fla.: Krieger, 1994.

Swenson, Loyd S., Jr., James M. Grimwood, and Charles C. Alexander. *This New Ocean: A History of Project Mercury*. Washington D.C.: NASA, 1966.

Trento, Joseph J., with Susan B. Trento. *Prescription for Disaster: From the Glory of Apollo to the Betrayal of the Shuttle*. New York: Crown, 1987.

Van Dyke, Vernon. *Pride and Power: The Rationale of the Space Program*. Urbana: University of Illinois Press, 1964.

Vietor, Richard H.K. *Contrived Competition: Regulation and Deregulation in America*. Boston: Belknap Press, 1994.

Webb, David C. "A Current Perspective On Space Commercialization: Removing Barriers to Opportunity." Washington, D.C.: Aerospace Industries Association of America, April 1985.

Whalen, David J. "Billion Dollar Technology: A Short Historical Overview of the Origins of Communications Satellite Technology, 1945–1965." In *Beyond the Ionosphere: Fifty Years of Satellite Communications*, edited by Andrew J. Butrica, 95–127. Washington, D.C.: NASA History Office (1997).

Wheelon, Albert D. "The Future of the Unmanned Space Program." In *Space Policy Alternatives*, edited by Radford Byerly, Jr., 13–34. Boulder, Colo.: Westview Press, 1992.

Wilks, Stephen, and Maurice Wright, eds. *Comparative Government-Industry Relations: Western Europe, the United States, and Japan*. Oxford: Clarendon Press, 1987.

Woodcock, Gordon R. *Space Stations and Platforms*. Malabar, Fla.: Orbit, 1986.

U.S. GOVERNMENT SOURCES
I. Congressional Hearings

Note: The House Committee with authority over NASA was created as the Select Committee on Astronautics and Space Exploration, in response to Sputnik. In 1959 it was converted into a standing committee with the name Committee on Science and Astronautics. In 1975 it was renamed the Committee on Science and Technology, in 1987, the Committee on Science, Space, and Technology, and in 1995, the Committee on Science.

House Committee on Appropriations. "Department of Defense Appropriations for 1985, Part 6." 98th Cong., 2d sess., May 22, 1984.

House Committee on Government Operations. "Cost, Justification, and Benefits of NASA's Space Station." 102d Cong., 1st sess., May 1, 1991.

House Committee on Science. "The X-33 Reusable Launch Vehicle: A New Way of Doing Business?" 104th Cong., 1st sess., November 1, 1995.

———. "NASA Procurement in the Earth-Space Economy." 104th Cong., 1st sess. November 8, 1995.

House Committee on Science and Astronautics. "Investigation into Apollo 204 Accident." 90th Cong., 1st sess., April 10 . . . , 1967.

House Committee on Science and Technology. "Space Industrialization." 95th Cong., 1st sess., September 29, 1977.

———. "Communications Research and Development." 96th Cong., 2d sess., May 1980.

———."NASA Space Communications Program," 97th Cong., 1st sess., July 8, 9, 1981.

———. "Future Space Programs: 1981." 97th Cong., 1st sess., July 8, 9, 1981, September 21–23, 1981.

———. "The Need for a Fifth Shuttle Orbiter." 97th Cong., 2d sess., June 15, 1982.

———. "National Space Policy." 97th Cong., 2d sess., August 4, 1982.

———. "Space Commercialization." 98th Cong., 1st sess., May 3–5, 1983.

———. "NASA's Space Station Activities." 98th Cong., 1st sess., August 2, 1983.

———. "Initiatives to Promote Space Commercialization." 98th Cong., 2d sess., June 19, 1984.

———. "Space Commercialization: 1985." 99th Cong., 1st sess., June 18–October 31, 1985.

———. "Assured Access to Space: 1986." 99th Cong., 2d sess., February 26–August 14, 1986.

———. "1987 NASA Authorization." 99th Cong., 2d sess., February–April 1986.

———. "1987 NASA Authorization, Vol 2." 99th Cong., 2d sess., March 20, 1986.

House Committee on Science, Space, and Technology. "Proposed Space Station Freedom Program Revisions." 101st Cong., 1st sess., October 31, 1989.

———. "Commercial Space Launch Act Implementation." 101st Cong., 1st sess., November 9, 1989.

———. "Review and Implementation of the Report of the Advisory Committee on the Future of the U.S. Space Program." 102d Cong., 1st sess., January 3, 1991.

———. "The Future of the U.S. Space Industrial Base." 103d Cong., 1st sess., February 1993.

———. "NASA's Commercial Space Programs." 103d Cong., 1st sess., October 20, 1993.

———. "National Space Transportation Policy." 103d Cong., 2d sess., September 20, 1994.

House Select Committee on Astronautics and Space Exploration. "Astronautics and Space Exploration." 85th Cong., 2d sess., April, May 1958.

Joint Economic Committee, "The Aerospace Industry." 102d Cong., 1st and 2d sess., December 3, 1991 and February 27, 1992.

Senate Committee on Aeronautical and Space Sciences. "Apollo Accident." 90th Cong., 1st sess., April 13 and 17, 1967.

———. "Space Shuttle Payloads." Part I, 93d Cong., 1st sess., October 30, 1973.

Senate Committee on Armed Services. "Inquiry Into Satellite and Missile Programs." 85th Cong., 1st and 2d sess., November 1957–January 1958.

———. "Air Force Space Launch Policy and Plans." 100th Cong., 1st sess., October 6, 1987.

Senate Committee on Commerce, Science, and Transportation. "Commercial Space Launch Act." 98th Cong. 1st sess., September 6, 1984.

———. "Commercial Space Opportunities." 100th Cong., 1st sess., October 5, 1987.

————. "The NASA Space Shuttle and the Reusable Launch Vehicle Programs." 104th Cong., 1st sess., May 16, 1995.

II. Other Government Publications

Congressional Budget Office. David H. Moore, "Setting Space Transportation Policy for the 1990s: A Special Study." Washington, D.C., Congressional Budget Office, October 1986.

————. *The NASA Program in the 1990s and Beyond.* Washington, D.C.: Congressional Budget Office, 1988.

Congressional Office of Technology Assessment. *SDI: Technology, Survivability, and Software,* OTA-ISC-353. Washington, D.C.: Government Printing Office, May 1988.

————. *Access to Space: The Future of U.S. Space Transportation Systems.* OTA-ISC-415. Washington D.C.: Government Printing Office, April 1990.

————. *The National Space Transportation Policy: Issues for Congress.* OTA-ISS-620. Washington D.C.: Government Printing Office, May 1995.

Congressional Research Service of the Library of Congress. "Policy and Legal Issues Involved in the Commercialization of Space." Report for the Senate Committee on Commerce, Science, and Transportation. Washington, D.C.: Government Printing Office, September 23, 1983.

————. "NASA's Advanced Communications Technology Satellite (ACTS) Program in Light of the Hughes Filing," by Marcia S. Smith. Washington D.C.: Government Printing Office, March 2, 1984.

————. "Commercial Space Launch Services: The U.S. Competitive Position." Report for the House Committee on Science, Space, and Technology. Washington D.C.: Government Printing Office, November 1991.

————. "U.S. Commercial Space Activities," by Tony Reichhardt. Washington D.C.: Government Printing Office, February 1, 1992.

Department Of Commerce, *Space Commerce: An Industry Assessment,* May 1988.

Index

The Library of Congress has cataloged the
hardcover edition of this book as follows:

Bromberg, Joan Lisa.
NASA and the space industry / Joan Lisa Bromberg
p.   cm. — (New series in NASA history)
Includes bibliographical references and index.
ISBN 0-8018-6050-4 (alk. paper)
1. United States. National Aeronautics and Space Administration—
History.   2. Aerospace industries—Government policy—United
States—History.   I. Title.   II. Series.
TL521.312.B76   1999
338.4′76291′0973—dc21   98-44795

ISBN 0-8018-6532-8 (pbk.)